Wavelets and Multiscale Signal Processing

APPLIED MATHEMATICS AND
MATHEMATICAL COMPUTATION

Editors

R.J. Knops, K.W. Morton

Text and monographs at graduate and research level covering a wide variety of topics of current research interest in modern and traditional applied mathematics, in numerical analysis, and computation.

1. Introduction to the Thermodynamics of Solids *J.L. Ericksen* (1991)

2. Order Stars *A. Iserles and S.P. Norsett* (1991)

3. Material Inhomogeneities in Elasticity *G. Maugin* (1993)

4. Bivectors and Waves in Mechanics and Optics *Ph. Boulanger and M. Hayes* (1993)

5. Mathematical Modelling of Inelastic Deformation *J.F. Besseling and E. Van der Geissen* (1993)

6. Vortex Structures in a Stratified Fluid: Order from chaos *Sergey I. Voropayer and Yakov D. Afanasyev* (1994)

7. Numerical Hamiltonian Problems *J.M. Sanz-Serna and M.P. Calvo* (1994)

8. Variational Theories for Liquid Crystals *Epifanio G. Virga* (1994)

9. Asymptotic Treatment of Differential Equations *Adelina Georgescu* (1995)

10. Plasma Physics Theory *A. Sitenko and V. Malnev* (1995)

11. Wavelets and Multiscale Signal Processing *Albert Cohen and Robert D. Ryan* (1995)

(Full details concerning this series, and more information on titles in preparation are available from the publisher.)

Wavelets and Multiscale Signal Processing

Albert Cohen

University of Paris VI

Robert D. Ryan

Mathematician and Translator
Paris, France

CHAPMAN & HALL

London · Glasgow · Weinheim · New York · Tokyo · Melbourne · Madras

Published by Chapman & Hall, 2–6 Boundary Row, London SE1 8HN, UK

Chapman & Hall, 2–6 Boundary Row, London SE1 8HN, UK

Blackie Academic & Professional, Wester Cleddens Road, Bishopbriggs, Glasgow G64 2NZ, UK

Chapman & Hall GmbH, Pappelallee 3, 69469 Weinheim, Germany

Chapman & Hall USA, 115 Fifth Avenue, New York, NY 10003, USA

Chapman & Hall Japan, ITP-Japan, Kyowa Building, 3F, 2-2-1 Hirakawacho, Chiyoda-ku, Tokyo 102, Japan

Chapman & Hall Australia, 102 Dodds Street, South Melbourne, Victoria 3205, Australia

Chapman & Hall India, R. Seshadri, 32 Second Main Road, CIT East, Madras 600 035, India

English language edition 1995

© 1995 Masson, Editeur, Paris

Original French language edition – Ondelettes et Traitement Numérique du Signal
© 1992, Masson, Paris.

Printed in England by Clays Ltd., St Ives plc

ISBN 0 412 57590 6

Apart from any fair dealing for the purposes of research or private study, or criticism or review, as permitted under the UK Copyright Designs and Patents Act, 1988, this publication may not be reproduced, stored, or transmitted, in any form or by any means, without the prior permission in writing of the publishers, or in the case of reprographic reproduction only in accordance with the terms of the licences issued by the Copyright Licensing Agency in the UK, or in accordance with the terms of licences issued by the appropriate Reproduction Rights Organization outside the UK. Enquiries concerning reproduction outside the terms stated here should be sent to the publishers at the London address printed on this page.

The publisher makes no representation, express or implied, with regard to the accuracy of the information contained in this book and cannot accept any legal responsibility or liability for any errors or omissions that may be made.

A catalogue record for this book is available from the British Library

∞ Printed on permanent acid-free text paper, manufactured in accordance with ANSI/NISO Z39.48-1992 and ANSI/NISO Z39.48-1984 (Permanence of Paper).

Contents

Foreword to the English Edition	**vii**
Introduction	**1**
1 Multiresolution analysis	**7**
1.1 Introduction	7
1.2 The continuous point of view	8
1.2.1 Multiresolution analyses and wavelets	8
1.2.2 The scaling function	11
1.2.3 Conjugate quadrature filters	16
1.3 The discrete point of view	20
1.3.1 The FWT algorithm	20
1.3.2 Multiresolution analyses of $l^2(\mathbb{Z})$	24
1.3.3 Orthonormal wavelet bases for $l^2(\mathbb{Z})$	27
1.4 The multivariate case	33
1.5 Conclusions	34
2 Wavelets and conjugate quadrature filters	**37**
2.1 Introduction	37
2.2 The general case	38
2.3 The finite case	46
2.4 Wavelets with compact support	53
2.5 Action of the FWT on oscillating signals	58
3 The regularity of scaling functions and wavelets	**63**
3.1 Introduction	63
3.2 Regularity and oscillation	64
3.3 The subdivision algorithms	68
3.4 Spectral estimates of the regularity	76
3.4.1 A condition sufficient for regularity	77
3.4.2 The critical exponent of a CQF	79

	3.4.3	The decay of $\hat{\varphi}$	80
	3.5	Estimation of the L^p-Sobolev exponent	84
	3.6	Applications	88
	3.6.1	Estimating the critical exponent	88
	3.6.2	The behavior for large values of N	93

4 Biorthogonal wavelet bases 97
4.1 Introduction 97
4.2 General principles of subband coding 99
4.3 Unconditional biorthogonal wavelet bases 103
4.4 Dual filters and biorthogonal Riesz bases 126
4.5 Examples and applications 144
    4.5.1 Burt and Adelson's filters 144
    4.5.2 Spline filters 151

5 Stochastic processes 165
5.1 Introduction 165
5.2 Linear approximation 169
5.3 Linear approximation of images 172
5.4 Approximation and compression of real images 174
5.5 Piecewise stationary processes 177
5.6 Non-linear approximation 181

A Quasi-analytic wavelet bases 199

B Multivariate constructions 207

C Multiscale unconditional bases 215

D Notation 225

References 229

Author index 233

Subject index 235

Foreword to the English Edition

This book is based on Albert Cohen's doctoral thesis, which was written at the University of Paris-Dauphine under the guidance of Professor Yves Meyer. It was published in 1992 by Masson, Paris. Since this first publication, research on wavelets and multiscale methods has progressed significantly, and interest in these fields has extended to a large scientific community. In spite of this progress, it was clear to us that the relations between filter banks and wavelet bases constitute a consistent and indispensable core for the theory and applications of wavelets and that it was useful to present these relations in detail. Our main concern in this edition has been to keep this focus on the core material, while at the same time providing 'pointers' to the latest results and related developments.

The major results presented here are from Cohen's thesis and from subsequent work done by Cohen and his collaborators. This English edition was translated by Robert D. Ryan. Extensive revisions were made jointly by Cohen and Ryan. These include revisions in Chapter 3, a new Section 4.3, and a rewritten Chapter 5. Most of the figures and plates have been redone to reflect changes in the text. In addition, an effort has been made to elaborate a number of the proofs to make the arguments more readily available to students and to others who are not experts in the field.

Albert Cohen
Robert D. Ryan
Paris, March 1995

Introduction

Joseph Fourier announced in 1807 that any 2π-periodic function could be represented as a series of sines and cosines. Since then, the spectral analysis of functions using Fourier series and integrals has been the source of numerous mathematical problems. Examples include the efforts to characterize function spaces in terms of their spectral properties and the study of operators on these spaces. In general, the problems encountered in these programs are related to the fact that it is impossible to describe the local properties of functions in terms of their spectral properties, which can be viewed as an expression of the Heisenberg uncertainty principle. Other problems arise because Fourier series may diverge in many of the usual function spaces.

Similar difficulties appear in the more applied areas of analog and digital signal processing, particularly when the signals represent non-stationary phenomena. Engineers, as well as mathematicians, have investigated and tested analytic methods that were better adapted to their problems and that attempted to avoid the difficulties inherent in classical Fourier analysis.

Some of these methods are very close to spectral analysis. For example, Dennis Gabor introduced a 'sliding-window' technique. He first multiplied the signal by the 'window,' for which he used a Gaussian function g, and he then calculated the Fourier transform. Thus, the analysing function is

$$g_{a,b}(x) = e^{iax} g(x-b), \qquad a, b \in \mathbb{R}.$$

This is a 'time–frequency' method, and while these methods are useful for certain applications, this analysing function has the disadvantage that the spatial resolution is limited by the fixed size of the Gaussian envelope.

In the early 1980s, Jean Morlet, who is a geophysicist, had the idea to base an analysis on one function h that would be well localized in both time and frequency. This function, which Morlet called a 'wavelet,' was then dilated and translated to form a family of analysing functions. These are normalized as follows:

$$h_{a,b}(x) = a^{-1/2} h\left(\frac{x-b}{a}\right), \qquad a \in \mathbb{R}^*, \, b \in \mathbb{R}.$$

With this technique, small values of the parameter a provide a local analysis of f, and the function of two variables,

$$Sf(a,b) = \langle f, h_{a,b} \rangle = \int a^{-1/2} f(x) \overline{h\left(\frac{x-b}{a}\right)} \, dx,$$

is a time-scale representation called a 'wavelet transform.' If the wavelet h satisfies the admissibility condition,

$$C_h = 2\pi \int \frac{|\hat{h}(\omega)|^2}{|\omega|} \, d\omega < +\infty,$$

(which means that h oscillates since $\hat{h}(0) = \int h(x)dx = 0$), then the transform can be inverted with the reconstruction formula

$$f(x) = \frac{1}{C_h} \iint Sf(a,b) \, h_{a,b}(x) \, \frac{da \, db}{a^2}.$$

An interesting property of this formula is that it converges in many function spaces where the Fourier transform fails to do so.

In 1985, Yves Meyer discovered, by carefully choosing the analysing function and by taking $a = 2^{-j}$ and $b = 2^{-j}k$ ($j, k \in \mathbb{Z}$) as discrete values for the parameters a and b, that one could obtain orthonormal bases for $L^2(\mathbb{R})$ of the type

$$\psi_{j,k}(x) = 2^{j/2} \psi(2^j x - k), \qquad j, k \in \mathbb{Z}, \tag{0.1}$$

and that the expression,

$$f = \sum_{j,k \in \mathbb{Z}} \langle f, \psi_{j,k} \rangle \, \psi_{j,k},$$

for decomposing a function in these orthonormal wavelets converged in many function spaces.

A particular example of orthonormal wavelets has been known since the beginning of the century, namely, the system introduced by Alfred Haar (1910). Unfortunately, the Haar wavelets are discontinuous and, consequently, poorly localized in frequency.

Stéphane Mallat made a decisive step in the theory of wavelets in 1987 when he proposed a fast algorithm for the computation of wavelet coefficients. He identified this algorithm with pyramidal schemes that decompose signals into subbands. These techniques were developed by several engineers in the 1970s (see Esteban and Galand (1977)) to reduce quantization noise. The framework that unifies these algorithms and the theory of wavelets is the concept of a multiresolution analysis. Briefly, a multiresolution analysis is an increasing sequence of closed, nested subspaces $\{V_j\}_{j\in\mathbb{Z}}$ that tends to $L^2(\mathbb{R})$ as j increases. One assumes, in addition, that V_j is obtained from V_{j+1} by a dilation of factor 2 and that there exists a function φ whose translates $\{\varphi(x-k)\}_{k\in\mathbb{Z}}$ form an orthonormal basis for V_0. It follows that φ satisfies a relation of the form

$$\varphi(x) = 2 \sum h_n \, \varphi(2x - n). \qquad (0.2)$$

This 'two-scale equation' plays an essential role in the theory of wavelet bases.

In the first place, by writing

$$\psi(x) = 2 \sum (-1)^n \, \overline{h_{1-n}} \, \varphi(2x - n), \qquad (0.3)$$

one generates a function whose translates $\{\psi(x-k)\}_{k\in\mathbb{Z}}$ form an orthonormal basis for the orthogonal complement, W_0, of V_0 in V_1. It follows that the complete family $\{\psi_{j,k}\}_{j,k\in\mathbb{Z}}$, as defined in (0.1), forms a Hilbert basis for $L^2(\mathbb{R})$. All of the currently known wavelet bases are obtained this way.

Thus the 'scaling function' φ, which is sometimes called the father wavelet, plays a key role in this theory. Linear combinations of its translates provide the 'V_0-approximation' for functions f in $L^2(\mathbb{R})$, while the translates of the function ψ carry the details that allow the approximations to be refined to the next scale, that is, to the 'V_1-approximation.'

On the other hand, the algorithms identified by Stéphane Mallat use discrete, sampled data that are analysed as if they were the 'approximation coefficients,' $s_k = \langle f, \varphi(\cdot - k)\rangle$, at the scale $j = 0$. As a consequence of the relations (0.2) and (0.3), Mallat's algorithms depend essentially on the coefficients h_n rather than on the functions φ and ψ. In turn, these coefficients are identified with the impulse response of a 'conjugate quadrature filter' (CQF) whose transfer function satisfies the relation

$$|m_0(\omega)|^2 + |m_0(\omega + \pi)|^2 = 1.$$

These filters were introduced by Smith and Barnwell (1986). Their complete classification, when the coefficients h_n are real and finite in number, was given by Ingrid Daubechies (1988) in the context of constructing wavelet bases with compact support.

Based on these results, it seemed natural to ask what are the contributions from the theory of wavelets to the field of subband filtering that can be investigated without being preoccupied with the underlying functional analysis? This book, which is based on work that started when the Fast Wavelet Transform (FWT) had just appeared, attempts to provide some answers to this question.

The theory of multiresolution analyses is developed in Chapter 1. We distinguish between the continuous and discrete points of view, which is to say, the analyses of the spaces $L^2(\mathbb{R})$ and $l^2(\mathbb{Z})$, and we relate the development to the FWT algorithm.

Chapter 2 is devoted to exploring the equivalence between the concept of a conjugate quadrature filter and that of a multiresolution analysis. This is an interesting problem because it can be treated by radically different approaches: arithmetic, algebraic, or algorithmic.

In Chapter 3, we study the regularity of wavelets in terms of the properties of their associated filters. We also examine the significance of regularity for digital signal processing.

An important generalization is developed in Chapter 4. Starting with a class of filters that is more general than the CQFs, we construct biorthogonal wavelet bases. Specifically, we construct two functions ψ and $\tilde{\psi}$ such that, for any $f \in L^2(\mathbb{R})$,

$$f = \sum_{j,k \in \mathbb{Z}} \langle f, \tilde{\psi}_{j,k} \rangle \psi_{j,k}.$$

In this case, there is clearly a contribution from digital signal processing to functional analysis.

Chapter 5 is devoted to the multiresolution analysis of stochastic processes in connection with signal and image compression. Wavelets are potentially useful in this context because, unlike complex exponentials, wavelets oscillate and are well localized. Thus, the wavelet coefficients that correspond to locations where a signal or image varies slowly are, *a priori*, very small, and one hopes to obtain an economical representation of the pertinent information.

We have included in three appendices further mathematical development of material selected from the first four chapters.

This book is based essentially on work from Albert Cohen's thesis, written under the direction of Professor Yves Meyer at Centre de Recherches en Mathématiques de la Décision (CEREMADE) at the University of Paris–Dauphine, and on subsequent research by Cohen and his collaborators, in particular, Ingrid Daubechies. Several articles, written between September 1988 and June 1990, have been recast in this book to provide a coherent presentation, as opposed to a collection of papers. As a result, the five chapters cannot easily be read independently. The concepts and results introduced in the first chapter are used throughout the rest of the book. We feel that readers will also benefit by taking Chapters 2, 3, and 4 in order. Appendix D contains the notation that we use throughout the book. We suggest that readers glance at this before beginning Chapter 1.

CHAPTER 1

Multiresolution analysis

1.1 Introduction

The purpose of this chapter is to present in parallel two methods of multiscale signal processing: the first is analog and the second is digital. From the mathematical point of view, this means that we will make a precise distinction between the analysis of functions in $L^2(\mathbb{R})$ and the analysis of sequences in $l^2(\mathbb{Z})$.

The multiscale approach in functional analysis appeared at the beginning of the century in work by Haar, Franklin, and Littlewood and Paley. The purpose was to deal with problems that were not resolved by the Fourier transform, such as the study of the regularity and other local properties of a function. One of the sequels of these efforts is the theory of orthonormal wavelets that was developed by Yves Meyer beginning in 1985.

In the more applied areas of signal processing for speech and images, electrical engineers were working on pyramidal representations of digital signals since the 1960s. The work of Esteban and Galand, and later work by Smith and Barnwell and by Burt and Adelson, provided various digital filtering schemes that can be used to produce multiscale decompositions of discrete data.

The multiresolution analyses that were introduced by Stéphane Mallat and Yves Meyer in 1987 are, as we will see, situated at the intersection of these two developments (see Meyer (1990)). Thus, an important problem is to clarify the exact connections between the two aspects of this theory. This is the objective of the next two sections of this chapter.

Our presentation will be restricted to one variable. The constructions of multiresolution analyses for $L^2(\mathbb{R}^n)$ are mentioned later in this chapter because they will be used in Chapter 5. They will be further developed in Appendix B. For most of the questions that we will pose, these constructions only add cumbersome notation because the essential issues are the same in $L^2(\mathbb{R})$ and in $L^2(\mathbb{R}^n)$.

1.2 The continuous point of view

1.2.1 Multiresolution analyses and wavelets

By definition, a multiresolution analysis is a sequence $\{V_j\}_{j \in \mathbb{Z}}$ of closed subspaces of $L^2(\mathbb{R})$ that satisfy the following properties:

$$V_j \subset V_{j+1} \tag{1.1}$$

$$f(x) \in V_j \iff f(2x) \in V_{j+1} \tag{1.2}$$

$$\bigcap_{j \in \mathbb{Z}} V_j = 0 \tag{1.3}$$

$$\bigcup_{j \in \mathbb{Z}} V_j \text{ is dense in } L^2(\mathbb{R}) \tag{1.4}$$

There exists a function $g(x)$ in V_0 such that $\{g(x-k)\}_{k \in \mathbb{Z}}$ is a Riesz basis for V_0. $\tag{1.5}$

Recall that a family $\{e_i\}$ is a Riesz basis for a Hilbert space H if and only if the following two properties are satisfied:
- The finite linear combinations $\sum \beta_i e_i$ are dense in H.
- There exist two strictly positive constants C_1 and C_2 such that for all sequences $\{\beta_i\}$, one has

$$C_1 \sum |\beta_i|^2 \leq \left\| \sum \beta_i e_i \right\|^2 \leq C_2 \sum |\beta_i|^2. \tag{1.6}$$

REMARKS
- The family $\{g(x-k)\}_{k \in \mathbb{Z}}$ is also called an unconditional basis since, for all sequences $\{s_k\}$ in $l^2(\mathbb{Z})$, the sum $\sum_{k \in \mathbb{Z}} s_k \, g(x-k)$ exists in V_0 and does not depend on the order of summation. Thus, the function g defines a sampling operator between V_0 and $l^2(\mathbb{Z})$, which is bijective, continuous, and commutes with integer shifts.
- Properties (1.3) and (1.4) can also be expressed in terms of the orthogonal projections P_j on the spaces V_j. The equivalent conditions are, respectively, that for all f in $L^2(\mathbb{R})$,

$$\|P_j f\| \xrightarrow{j \to -\infty} 0, \tag{1.7}$$

$$\|P_j f - f\| \xrightarrow{j \to +\infty} 0. \tag{1.8}$$

- An immediate consequence of properties (1.2) and (1.5) is that

$$f(x) \in V_j \iff f(x - 2^{-j}k) \in V_j. \tag{1.9}$$

For this reason, V_j is called a 'shift-invariant space.'

- The projection $P_j f$ can be considered as an approximation of f at the scale 2^{-j}. Note that the shift-invariant spaces are well known in approximation theory. A classic example is given by the spline functions with equally spaced knots. These are the functions in $\mathcal{C}^{N-1}$ whose restrictions to the intervals $[kh, (k+1)h]$, $k \in \mathbb{Z}$, are polynomials of degree N. By taking $h = 2^{-j}$ and requiring that the functions are also in $L^2(\mathbb{R})$, we immediately obtain a multiresolution analysis. In this case, the space V_0 is generated by the basic spline $g_N(x) = (*)^{N+1} \mathbf{1}_{[0,1]}(x)$. We refer to de Boor (1978) for a general introduction to spline functions.

- The linear combination $f(x) = \sum_{k \in \mathbb{Z}} \alpha_k \, g(x-k)$ can also be expressed in the spectral domain as

$$\hat{f}(\omega) = m(\omega)\hat{g}(\omega) = \left(\sum_{k \in \mathbb{Z}} \alpha_k e^{-ik\omega}\right)\hat{g}(\omega), \qquad (1.10)$$

where the function $m(\omega)$ is 2π-periodic.

This relation, and the fact that $\frac{1}{2\pi}\int_0^{2\pi} |m(\omega)|^2 \, d\omega = \sum |\alpha_i|^2$, allows us to interpret the property of an unconditional basis expressed by (1.6) in the spectral domain, which means that the norm $\|m(\omega)\|$ in $L^2[0, 2\pi]$ is equivalent to $\left(\int_{\mathbb{R}} |m(\omega)|^2 |\hat{g}(\omega)|^2 \, d\omega\right)^{1/2}$. Explicitly, (1.6) implies that

$$\begin{aligned} C_1 \int_0^{2\pi} |m(\omega)|^2 \, d\omega &\leq \int_0^{2\pi} |m(\omega)|^2 \left[\sum_{l \in \mathbb{Z}} |\hat{g}(\omega + 2l\pi)|^2\right] d\omega \\ &\leq C_2 \int_0^{2\pi} |m(\omega)|^2 \, d\omega \end{aligned} \qquad (1.11)$$

for all sequences $\{\alpha_k\}$ in $l^2(\mathbb{Z})$, or equivalently, for all m in $L^2[0, 2\pi]$. Thus, (1.11) holds if and only if

$$0 < C_1 \leq \sum_{l \in \mathbb{Z}} |\hat{g}(\omega + 2l\pi)|^2 \leq C_2 \qquad (1.12)$$

for almost all ω. (Note that $\hat{g}(\omega)$ is defined for almost all ω and that the sum $\sum_{l \in \mathbb{Z}} |\hat{g}(\omega + 2l\pi)|^2$ is defined and, by (1.12), finite for almost all ω.)

Based on these observations, we can construct, without making further assumptions, a function φ in $L^2(\mathbb{R})$ such that the family $\{\varphi(x-k)\}_{k \in \mathbb{Z}}$ is an orthonormal basis for the space V_0.

This function is defined by

$$\hat{\varphi}(\omega) = \left(\sum_{l\in\mathbb{Z}} |\hat{g}(\omega+2l\pi)|^2\right)^{-1/2} \hat{g}(\omega). \qquad (1.13)$$

By summing, we have

$$\sum_{l\in\mathbb{Z}} |\hat{\varphi}(\omega+2l\pi)|^2 = 1, \qquad (1.14)$$

which can also be written as

$$\frac{1}{2\pi} \int |\hat{\varphi}(\omega)|^2 \, e^{ik\omega} \, d\omega = \delta_{0,k}. \qquad (1.15)$$

Finally, the orthogonality property,

$$\langle \varphi, \varphi(\cdot - k) \rangle = \int \varphi(x)\,\overline{\varphi(x-k)}\,dx = \delta_{0,k}, \qquad (1.16)$$

follows directly from (1.15), Parseval's formula, and the relation $\hat{\varphi}(\omega)e^{-ik\omega} = \widehat{\varphi(\cdot-k)}(\omega)$.

The function φ plays an essential role in everything that follows. It is not a wavelet, however; φ is called the scaling function or sometimes the 'father wavelet.'

The orthonormal wavelets bases can be introduced as follows: We consider W_j, the orthogonal complement of V_j in V_{j+1}, that is, $V_{j+1} = V_j \stackrel{\perp}{\oplus} W_j$. W_j represents the additional information that is necessary to pass from the approximation at the scale 2^j to the approximation at the finer scale 2^{j+1}.

Suppose for the moment, as in the case of V_0, that there is a function ψ in W_0 such that the family $\{\psi(x-k)\}_{k\in\mathbb{Z}}$ is an orthonormal basis for W_0. (Later, we will see that the wavelet ψ can be constructed directly from the function φ.) It is then clear, by a change of scale, that the families $\{\psi_{j,k}\}_{k\in\mathbb{Z}} = \{2^{j/2}\,\psi(2^j x - k)\}_{k\in\mathbb{Z}}$ are orthonormal bases for the spaces W_j. Furthermore, from properties (1.3) and (1.4), we have

$$\stackrel{\perp}{\bigoplus_{j\in\mathbb{Z}}} W_j = L^2(\mathbb{R}). \qquad (1.17)$$

The family $\{\psi_{j,k}\}_{j,k\in\mathbb{Z}}$ thus forms a Hilbert basis for the entire space $L^2(\mathbb{R})$, and this is called a wavelet basis.

A simple example — which is, however, historically important — is given when V_0 is the space of functions that are constant on the intervals $[m, m+1]$, $m \in \mathbb{Z}$. These are the splines of order $N = 0$. In this case, the scaling function φ is $\mathbf{1}_{[0,1)}$, and the wavelet ψ is defined by $\psi(x) = \mathbf{1}_{[0,1/2)}(x) - \mathbf{1}_{[1/2,1)}(x)$. This function generates the Haar system, which was introduced by Alfred Haar (1910). The theory of wavelets allows us to introduce less trivial basis functions that extend the work of Haar. These functions will oscillate, as do the Haar functions, but, unlike the Haar functions, modern wavelets can have prescribed regularity properties.

We are going to set aside the discussion of wavelets for the moment and investigate the scaling function in greater detail.

1.2.2 The scaling function

We first introduce a quite general setting for studying the properties of the scaling function.

Definition 1.1 *A multiresolution analysis is said to be 'localized' if and only if the function φ satisfies the following property:*

$$\int (1 + |x|)^m |\varphi(x)|^2 dx < +\infty \qquad \text{for all} \quad m \in \mathbb{N}. \qquad (1.18)$$

This also means that the function $\hat{\varphi}(\omega)$ belongs to all of the Sobolev spaces $H^m(\mathbb{R})$.

Here are several observations about this definition:

• It follows from (1.14) and (1.15) that the choice of φ is unique up to a multiplication of $\hat{\varphi}$ by a 2π-periodic function that is regular and unimodular.

• This definition may seem rather restrictive, but in practice it is not because the scaling functions that are generally considered either have compact support or decrease rapidly at infinity. In both of these case, (1.18) is satisfied.

• The property expressed by (1.18) cannot be used to characterize the behavior of the function φ at infinity in the usual terms of pointwise decay, such as 'decreasing rapidly at infinity.' One can, however, deduce bounds for the 'tails' of integrals of the sort

$$\int_{|x|>A} |\varphi(x)|^2 \, dx \leq \frac{C_m}{A^m} \qquad \text{for all} \quad m \in \mathbb{N}. \qquad (1.19)$$

• If (1.18) holds, then an application of Schwarz's inequality shows that φ is in $L^1(\mathbb{R})$ and that (1.18) is also true with $|\varphi(x)|^2$ replaced by $|\varphi(x)|$. Thus, (1.19) also holds with $|\varphi(x)|^2$ replaced by

$|\varphi(x)|$. These kinds of estimates will be very useful for establishing many of the general results about multiresolution analyses.

Finally, to emphasize the central role of the scaling function in all of the theory, we pose the following question:

Let φ be a function that satisfies (1.18) and such that $\{\varphi(x-k)\}_{k\in\mathbb{Z}}$ is an orthonormal family. If we then define V_j as the space generated by the basis $\{\varphi_{j,k}\}_{k\in\mathbb{Z}} = \{2^{j/2}\,\varphi(2^j x - k)\}_{k\in\mathbb{Z}}$, can one say that the sequence $\{V_j\}_{j\in\mathbb{Z}}$ forms a multiresolution analysis? It is clear by construction that properties (1.2) and (1.5) are satisfied, but what about the other three?

The following theorem clarifies this point.

Theorem 1.1 *Let φ be a function in $L^2(\mathbb{R})$ that satisfies properties (1.14) and (1.18) and let V_j be the closed subspaces generated by the families $\{\varphi_{j,k}\}_{k\in\mathbb{Z}}$. Then*

(a) (1.1) $\iff$ there exists a function $m_0(\omega)$, which is $\mathcal{C}^\infty$ and 2π-periodic, such that $\hat{\varphi}(2\omega) = m_0(\omega)\hat{\varphi}(\omega)$,

(b) (1.3) is always satisfied, and

(c) (1.4) $\iff |\hat{\varphi}(0)| = \left|\int \varphi(x)\,dx\right| = 1$.

Under these conditions, the function φ defines a localized multiresolution analysis.

Proof. (a) It is clear from property (1.1) that $\varphi(x/2)$ can be expressed in terms of the basis functions $\varphi(x-k)$. Indeed, we can write

$$\frac{1}{2}\varphi(x/2) = \sum_{k\in\mathbb{Z}} h_k\,\varphi(x-k), \qquad (1.20)$$

where the sequence $\{h_k\}_{k\in\mathbb{Z}}$ is in $l^2(\mathbb{Z})$. Taking the Fourier transform of both sides gives

$$\hat{\varphi}(2\omega) = m_0(\omega)\,\hat{\varphi}(\omega), \qquad (1.21)$$

where $m_0(\omega)$ is the discrete Fourier transform of $\{h_k\}_{k\in\mathbb{Z}}$. But because $\{\varphi_{0,k}\}_{k\in\mathbb{Z}}$ is assumed to be an orthonormal family (by (1.14)), we can also express the coefficients h_k by

$$h_k = \frac{1}{2}\int \varphi(x/2)\,\overline{\varphi(x-k)}\,dx. \qquad (1.22)$$

We estimate this integral by first separating it into two integrals: one for the values $|x| \leq \frac{|k|}{2}$ and the other for the values $|x| > \frac{|k|}{2}$.

THE CONTINUOUS POINT OF VIEW

Using Schwarz's inequality, we have

$$|h_k| \leq \frac{1}{\sqrt{2}} \left[\int_{|x| \leq \frac{|k|}{2}} |\varphi(x-k)|^2 \, dx \right]^{1/2} + \frac{1}{2} \left[\int_{|x| > \frac{|k|}{2}} |\varphi(x/2)|^2 \, dx \right]^{1/2}.$$

Straightforward manipulation of these integrals and two applications of the inequality (1.19) show that

$$|h_k| \leq C_m (1+|k|)^{-m} \qquad \text{for all } m \in \mathbb{N}. \tag{1.23}$$

(Note that here and elsewhere constants will change values without notice. In particular, the constants C_m in (1.19) are not the same as the constants C_m in (1.23).)

Thus the Fourier coefficients of $m_0(\omega)$ decrease rapidly at infinity, and this implies that the function $m_0(\omega)$ is in $\mathcal{C}^\infty(\mathbb{R})$. It is clear, in the other direction, that the existence of a $\mathcal{C}^\infty$, 2π-periodic, function $m_0(\omega)$, which satisfies (1.21), implies the inclusion $V_{-1} \subset V_0$. The general inclusion (1.1) follows by changing scale.

(b) We wish to show that, for each f in $L^2(\mathbb{R})$, the projections $P_j f$ of f on the subspaces V_j converge to 0 in norm when j tends to $-\infty$. For this, we use the fact that the step functions with compact support are dense in $L^2(\mathbb{R})$. Thus, for any $f \in L^2(\mathbb{R})$, we can write $f = g + h$ where $\|h\|$ is arbitrarily small and g is a finite linear combination of characteristic functions of bounded intervals. Then $\|P_j f\| \leq \|P_j g\| + \|h\|$ since $\|P_j\| = 1$.

Consequently, it is sufficient to show that $\|P_j \mathbf{1}_{[a,b]}\|$ tends to zero for each finite interval $[a,b] \subset \mathbb{R}$. We use the assumption that $\{\varphi_{j,k}\}_{k \in \mathbb{Z}}$ is an orthonormal basis for V_j and write

$$\begin{aligned}
\|P_j \mathbf{1}_{[a,b]}\|^2 &= \sum_{k \in \mathbb{Z}} |\langle \mathbf{1}_{[a,b]}, \varphi_{j,k} \rangle|^2 \\
&= 2^j \sum_{k \in \mathbb{Z}} \left| \int_a^b \overline{\varphi(2^j x - k)} \, dx \right|^2 \\
&= 2^{-j} \sum_{k \in \mathbb{Z}} \left| \int_{2^j a}^{2^j b} \overline{\varphi(x - k)} \, dx \right|^2.
\end{aligned}$$

Applying Schwarz's inequality to the last equation shows that

$$\|P_j \mathbf{1}_{[a,b]}\|^2 \leq (b-a) \sum_{k\in\mathbb{Z}} \int_{2^j a}^{2^j b} |\varphi(x-k)|^2 \, dx$$

$$= \int \sum_{k\in\mathbb{Z}} \mathbf{1}_{[2^j a+k,\, 2^j b+k]}(x) \, |\varphi(x)|^2 \, dx.$$

If $-j$ is sufficiently large, the intervals $\{[2^j a + k,\, 2^j b + k]\}_{k\in\mathbb{Z}}$ are disjoint and $\sum_{k\in\mathbb{Z}} \mathbf{1}_{[2^j a+k,\, 2^j b+k]}(x) \, |\varphi(x)|^2 \leq |\varphi(x)|^2$. For almost every $x \in \mathbb{R}$, $\sum_{k\in\mathbb{Z}} \mathbf{1}_{[2^j a+k,\, 2^j b+k]}(x) \downarrow 0$ when j tends to $-\infty$. Hence, by dominated convergence, the last integral tends to 0 as j tends to $-\infty$, and this proves (b).

Note that this proof does not use (1.18), and, in fact, we have shown that (1.3) is true whenever φ is in $L^2(\mathbb{R})$ and satisfies (1.14), which is the assumption that implies that the family $\{\varphi_{0,k}\}_{k\in\mathbb{Z}}$ is orthonormal.

(c) We wish to show that $\|P_j f - f\| \to 0$ for all f in $L^2(\mathbb{R})$ when j tends to $+\infty$ (which is equivalent to (1.4)) if and only if $|\hat{\varphi}(0)| = \left|\int \varphi(x) \, dx\right| = 1$.

As in (b), it is sufficient to consider $f = \mathbf{1}_{[a,b]}$. Furthermore, because the P_j are projections, we have

$$\|P_j f - f\|^2 = \|f\|^2 - \|P_j f\|^2, \tag{1.24}$$

and convergence in norm is equivalent to showing that $\|P_j \mathbf{1}_{[a,b]}\|^2$ tends to $b - a$ when j tends to $+\infty$.

Recall that condition (1.18) implies that φ is in $L^1(\mathbb{R})$ so that $|\hat{\varphi}(0)| = \left|\int \varphi(x) \, dx\right| = c$ is well defined. The discussion below will show that, under the assumptions (1.14) and (1.18), $\|P_j \mathbf{1}_{[a,b]}\|^2$ converges to $c^2(b-a)$ when j tends to $+\infty$. Thus, (c) is true if and only if $c = 1$.

We know that $\|P_j \mathbf{1}_{[a,b]}\|^2 = 2^{-j} \sum_{k\in\mathbb{Z}} |\int_{2^j a}^{2^j b} \varphi(x-k) \, dx|^2$. To estimate this expression, we separate the sum into three parts, writing $\|P_j \mathbf{1}_{[a,b]}\|^2 = A_j + B_j + C_j$, where

$$A_j = 2^{-j} \sum_{k \in \mathbb{Z}_{ja}} \left| \int_{2^j a}^{2^j b} \varphi(x-k) \, dx \right|^2,$$

$$\mathbb{Z}_{ja} = [2^j a + 2^{j/2},\, 2^j b - 2^{j/2}] \cap \mathbb{Z};$$

$$B_j = 2^{-j} \sum_{k \in \mathbb{Z}_{jb}} \left| \int_{2^j a}^{2^j b} \varphi(x-k) \, dx \right|^2,$$

$$\mathbb{Z}_{jb} = \mathbb{Z} \setminus [2^j a - 2^{j/2},\, 2^j b + 2^{j/2}];$$

and
$$C_j = 2^{-j} \sum_{k \in \mathbb{Z}_{jc}} \left| \int_{2^j a}^{2^j b} \varphi(x-k)\, dx \right|^2,$$
$$\mathbb{Z}_{jc} = \mathbb{Z} \setminus (\mathbb{Z}_{ja} \cup \mathbb{Z}_{jb}).$$

We estimate these three terms separately and differently, beginning with B_j. For this we have the following inequality:

$$B_j \leq 2^{-j} \sum_{k < 2^j a - 2^{j/2}} \left(\int_{2^j a - k}^{\infty} |\varphi(x)|\, dx \right)^2$$
$$+ 2^{-j} \sum_{k > 2^j b + 2^{j/2}} \left(\int_{-\infty}^{2^j b - k} |\varphi(x)|\, dx \right)^2.$$

The first sum is dominated by $\sum_{l=0}^{\infty} (\int_{2^{j/2}+l}^{\infty} |\varphi(x)|\, dx)^2$ and, similarly, the second sum is dominated by $\sum_{l=0}^{\infty} (\int_{-\infty}^{-2^{j/2}-l} |\varphi(x)|\, dx)^2$. We estimate these sums by using the inequality (1.19) applied to $|\varphi|$, which shows that

$$B_j \leq 2^{-j} C_m^2 \sum_{l=0}^{\infty} (2^{j/2} + l)^{-2m}$$

for all m. Since the sum on the right is bounded uniformly in j, this proves that B_j tends to 0 when j tends to $+\infty$.

Estimating C_j is straightforward because the sum is finite for each fixed j. We have

$$\begin{aligned}
C_j &\leq 2^{-j} \sum_{k \in \mathbb{Z}_{jc}} \left(\int_{2^j a}^{2^j b} |\varphi(x-k)|\, dx \right)^2 \\
&\leq 2^{-j} \sum_{k \in \mathbb{Z}_{jc}} \left(\int |\varphi(x)|\, dx \right)^2 \\
&= 2^{-j} \left(\int |\varphi(x)|\, dx \right)^2 (\mathrm{Card}(\mathbb{Z}_{jc})).
\end{aligned}$$

Since $(\mathrm{Card}(\mathbb{Z}_{jc})) \leq 2^{-j/2+2}$, we have

$$C_j \leq 2^{-j/2+2} \left(\int |\varphi(x)|\, dx \right)^2,$$

and this shows that C_j also tends to 0.

It remains to analyze A_j. For this, note that

$$\left| \int_{-\infty}^{\infty} \varphi(x-k)\,dx - \int_{2^j a}^{2^j b} \varphi(x-k)\,dx \right| \leq \int_{|x| \geq 2^{j/2}} |\varphi(x)|\,dx$$
$$\leq C_m\, 2^{-jm/2}$$

uniformly for $k \in \mathbb{Z}_{jc}$. This implies that the integrals in A_j tend to $|\int \varphi(x)\,dx| = c$ uniformly in k. Combined with the observation that $2^{-j}(\mathrm{Card}(\mathbb{Z}_{ja}))$ tends to $b - a$, this proves that A_j tends to $c^2(b-a)$ and that the same is true for $\|P_j \mathbf{1}_{[a,b]}\|^2$. Thus we have shown that (c) is true if and only if $c = 1$, and this completes the proof of Theorem 1.1. $\square$

Next, we are going to examine the nature of the function $m_0(\omega)$ that appeared in relation (1.21). This function will play a very important role in what follows.

1.2.3 Conjugate quadrature filters

The relation
$$\hat{\varphi}(2\omega) = m_0(\omega)\,\hat{\varphi}(\omega) \tag{1.25}$$
introduced the regular, 2π-periodic function m_0 that 'links' the two successive scales of approximation.

Recall that m_0 is the discrete Fourier transform of the sequence $\{h_k\}$ and that these numbers are the coordinates of $\frac{1}{2}\varphi(x/2)$ in the basis $\{\varphi(x-k)\}_{k \in \mathbb{Z}}$. In the language of signal processing, m_0 is the transfer function of a discrete filter with impulse response $\{h_k\}$. The term 'filter' will be justified shortly.

We can use the orthogonality of the sequence $\{\varphi(x-k)\}_{k \in \mathbb{Z}}$ to establish an important property of this filter. We insert relation (1.21) in the identity (1.14), which gives us

$$1 = \sum_{l \in \mathbb{Z}} |\hat{\varphi}(2\omega + 2l\pi)|^2$$
$$= \sum_{l \in \mathbb{Z}} |\hat{\varphi}(\omega + l\pi)|^2\, |m_0(\omega + l\pi)|^2$$
$$= \sum_{l \in \mathbb{Z}} |\hat{\varphi}(\omega + 2l\pi)|^2\, |m_0(\omega)|^2$$
$$+ \sum_{l \in \mathbb{Z}} |\hat{\varphi}(\omega + \pi + 2l\pi)|^2\, |m_0(\omega + \pi)|^2,$$

and finally
$$|m_0(\omega)|^2 + |m_0(\omega + \pi)|^2 = 1. \tag{1.26}$$
Since $|\hat{\varphi}(0)| = 1 \neq 0$,
$$m_0(0) = 1 \quad \text{and} \quad m_0(\pi) = 0. \tag{1.27}$$

The function m_0 is, in a weak sense, a 'low-pass filter' because the transfer function 'passes' the frequencies near $\omega = 0$ and 'cuts off' the frequencies near $\omega = \pi$. One can define the filter that is conjugate to m_0 by

$$m_1(\omega) = e^{-i\omega}\overline{m_0(\omega + \pi)} = \sum_{k \in \mathbb{Z}} g_k e^{-ik\omega} = \sum_{k \in \mathbb{Z}} (-1)^k \overline{h_{1-k}} e^{-ik\omega}.$$

A possible choice for the wavelet ψ is then given by the relation

$$\hat{\psi}(2\omega) = m_1(\omega)\,\hat{\varphi}(\omega). \tag{1.28}$$

We prove that $\{\psi(x-k)\}_{k \in \mathbb{Z}}$ is an orthonormal basis for W_0, the orthogonal complement of V_0 in V_1, by first showing that (1.14) holds for ψ. Thus write

$$\sum_{l \in \mathbb{Z}} |\hat{\psi}(\omega + 2l\pi)|^2 = \sum_{l \in \mathbb{Z}} |\hat{\varphi}(\frac{\omega}{2} + l\pi)|^2 |m_1(\frac{\omega}{2} + l\pi)|^2$$
$$= \sum_{l \in \mathbb{Z}} |\hat{\varphi}(\frac{\omega}{2} + 2l\pi)|^2 |m_1(\frac{\omega}{2})|^2$$
$$+ \sum_{l \in \mathbb{Z}} |\hat{\varphi}(\frac{\omega}{2} + \pi + 2l\pi)|^2 |m_1(\frac{\omega}{2} + \pi)|^2$$
$$= |m_1(\frac{\omega}{2})|^2 + |m_1(\frac{\omega}{2} + \pi)|^2$$
$$= 1.$$

This relation, combined with the same arguments that we used for the scaling function φ, shows that the family $\{\psi(x-k)\}_{k \in \mathbb{Z}}$ is orthonormal. A similar computation leads to

$$\sum_{l \in \mathbb{Z}} \hat{\psi}(\omega + 2l\pi)\overline{\hat{\varphi}(\omega + 2l\pi)} = 0,$$

which is equivalent to $\langle \psi(\cdot - k), \varphi(\cdot - l) \rangle = 0$ for all $k, l \in \mathbb{Z}$. Hence, V_0 and W_0 are orthogonal.

Finally, by using the properties of m_0 and m_1, it is easy to verify that

$$\hat{\varphi}(\frac{\omega}{2}) = \left[\overline{m_0(\frac{\omega}{2})} + \overline{m_0(\frac{\omega}{2}+\pi)}\right]\hat{\varphi}(\omega)$$
$$+ \left[\overline{m_1(\frac{\omega}{2})} + \overline{m_1(\frac{\omega}{2}+\pi)}\right]\hat{\psi}(\omega),$$

and

$$e^{-i\frac{\omega}{2}}\hat{\varphi}(\frac{\omega}{2}) = \left[e^{-i\frac{\omega}{2}}\overline{m_0(\frac{\omega}{2})} - e^{-i\frac{\omega}{2}}\overline{m_0(\frac{\omega}{2}+\pi)}\right]\hat{\varphi}(\omega)$$
$$+ \left[e^{-i\frac{\omega}{2}}\overline{m_1(\frac{\omega}{2})} - e^{-i\frac{\omega}{2}}\overline{m_1(\frac{\omega}{2}+\pi)}\right]\hat{\psi}(\omega).$$

We observe two things about these last equations: The left-hand members are, after dividing by 2, the Fourier transforms of $\varphi(2x)$ and $\varphi(2x-1)$, and the functions in brackets on the right-hand side are 2π-periodic and have the same regularity properties as m_0. This means that both $\varphi(2x)$ and $\varphi(2x-1)$ belong to the direct sum $V_0 \overset{\perp}{\oplus} W_0$. Since this subspace is invariant under integer shifts, it follows that $\varphi(2x-k) \in V_0 \overset{\perp}{\oplus} W_0$ for all $k \in \mathbb{Z}$. Thus, $V_1 = V_0 \overset{\perp}{\oplus} W_0$, which is what we wished to prove.

Note that the properties of m_0 and m_1 that we have used to prove this result can be summarized by stating that the matrix

$$\begin{pmatrix} m_0(\omega) & m_1(\omega) \\ m_0(\omega+\pi) & m_1(\omega+\pi) \end{pmatrix} \qquad (1.29)$$

is unitary for all $\omega \in \mathbb{R}$.

The filters m_0 and m_1 are called 'conjugate quadrature filters' and denoted by CQF. By an abuse of the language, we will call any $\mathcal{C}^\infty$, 2π-periodic function m_0 that satisfies (1.26) and (1.27) a CQF.

REMARKS

• Since the scaling function φ that generates the multiresolution analysis can be defined up to a multiplication of its Fourier transform $\hat{\varphi}$ by a 2π-periodic, unimodular, $\mathcal{C}^\infty$ function, we will assume that $\hat{\varphi}(0) = \int \varphi(x)\,dx = 1$.

• The first filters of this sort to appear in the signal processing literature were the 'quadrature mirror filters' that were introduced by Esteban and Galand (1977). These were later modified by Smith and Barnwell (1986) who were the first to use the form described by the relation (1.26).

THE CONTINUOUS POINT OF VIEW

• In all that follows, we will study only wavelet bases that are constructed from a multiresolution analysis. These are, in fact, the only ones that are currently known. Thus the spaces W_j can be interpreted as the details that are necessary to pass from the approximation in V_j to the finer approximation in V_{j+1}. The complete description of wavelet bases that are defined in full generality as translates and dilates of one 'mother' function has remained for some years an open problem. A recent result by P. G. Lemarié (1993) states that if ψ has compact support, then it is necessarily associated with a scaling function φ that also has compact support.

The importance of the function $m_0(\omega)$ will become fully apparent in the construction of the discrete multiresolution analyses. For the moment, we observe that the relation (1.21) can be iterated by changing scale, which shows that

$$\hat{\varphi}(\omega) = \prod_{k=1}^{n} m_0(2^{-k}\omega)\, \hat{\varphi}(2^{-n}\omega). \tag{1.30}$$

Because $\hat{\varphi}(0) = 1$ and $\hat{\varphi}$ is regular, $\hat{\varphi}(\omega)$ can be expressed by the infinite product

$$\hat{\varphi}(\omega) = \prod_{k=1}^{+\infty} m_0(2^{-k}\omega). \tag{1.31}$$

The regularity of the function m_0 at 0 ensures the pointwise convergence of the product.

In view of this formula and the results of Theorem 1.1, it appears that all of the theory that we have presented up to this point can be based on a conjugate quadrature filter (CQF). Thus, it is reasonable to ask if every CQF comes from a localized multiresolution analysis such as we have described. This topic will take on its full significance in the development of the discrete point of view, but we can immediately provide a negative answer to the question as posed.

Proposition 1.1 *For each function m_0 that comes from a localized multiresolution analysis one can create an infinite family of functions $\{m_{0,p}\}_{p \in \mathbb{N}^*}$ that are regular, 2π-periodic, and have all the properties of the CQFs, but that are not associated with multiresolution analyses of $L^2(\mathbb{R})$.*

Proof. We define the functions by

$$m_{0,p}(\omega) = m_0((2p+1)\omega). \tag{1.32}$$

It is clear that these filters satisfy (1.26) and (1.27).

If the $m_{0,p}$ were associated with multiresolution analyses, then the corresponding scaling functions φ_p could be computed by using the infinite product (1.31). Thus,

$$\hat{\varphi}_p(\omega) = \prod_{k=1}^{+\infty} m_{0,p}(2^{-k}\omega) = \hat{\varphi}((2p+1)\omega), \qquad (1.33)$$

where φ is the scaling function associated with m_0. If φ has norm 1 in $L^2(\mathbb{R})$, it is clear that φ_p has norm $1/\sqrt{2p+1}$. This means that the family $\{\varphi_p(x-k)\}_{k\in\mathbb{Z}}$ is not orthonormal, and proves the proposition. □

Notice also that $m_{0,p}(\omega) = 0$ if ω is an odd multiple of $\frac{\pi}{2p+1}$. Thus by (1.33), $\hat{\varphi}_p(\omega)$ vanishes at the even multiples of $\frac{\pi}{2p+1}$ (with the exception of 0) since at least one of the factors in the infinite product is zero. Consequently,

$$\sum_{l\in\mathbb{Z}} \left|\hat{\varphi}_p\left(\frac{2\pi}{2p+1} + 2l\pi\right)\right|^2 = 0, \qquad (1.34)$$

which implies that the family $\{\varphi_p(x-k)\}_{k\in\mathbb{Z}}$ cannot even be a Riesz basis.

Furthermore, this tells us that the vanishing of the transfer function at the points $\frac{(2k+1)\pi}{2p+1}$ is itself a contradiction to the existence of an underlying multiresolution analysis.

A simple example of this construction is given by dilating the Haar system. In this case, for $p = 1$ we have $m_{0,1}(\omega) = \frac{1+e^{-i3\omega}}{2}$. Then $\varphi_1 = \frac{1}{3}\mathbf{1}_{[0,3)}$, and its translates are clearly not orthonormal.

Stronger hypotheses, which are presented in the next chapter, will allow us to characterize those CQFs that are associated with multiresolution analyses. But at this point we take up the discrete theory in which the CQFs play a central role.

1.3 The discrete point of view

1.3.1 The FWT algorithm

The idea of developing an algorithm that allows one to obtain the wavelet coefficients of a signal gives rise to several comments. In the first place, the only data that can be processed by an algorithm are necessarily discrete, and, hence, we must abandon the functions of $L^2(\mathbb{R})$. An initial idea, which seems natural, is the following:

Sample the wavelets $\{\psi_{j,k}\}$ with a sampling step that corresponds to the sampling rate of the signal and do the signal analysis by computing the scalar products as series in the usual way, that is, by replacing the integrals with their Riemann sums.

Note that this method is used effectively to calculate the discrete Fourier transform. Here one samples the functions $e^{ik\omega}$ at points that are particularly well adapted to the calculation.

In the case of wavelets, such an idea is doomed to failure for four different reasons:

• Wavelets are elements of the space $L^2(\mathbb{R})$, and their point values have no *a priori* meaning. This poses a problem for sampling.

• Such a sampling would lead to numerical errors in the reconstruction of the signal. One would like to have a discrete analysis that also provides an exact reconstruction formula.

• For large values of the scale parameter j, it is clear that the sampling of the wavelet $\{\psi_{j,k}\}$ will be incompatible with Shannon's rules since the $\{\psi_{j,k}\}$ have oscillations that will not be detected by the fixed sampling rate.

• Finally, for small values of j, the support of the wavelet $\{\psi_{j,k}\}$ is so large that the cost of computing the coefficients becomes unrealistic.

It is thus necessary to abandon this idea, which works so well for the discrete Fourier transform. The principle of the FWT algorithm, which was developed by Stéphane Mallat, is completely different and can be summarized as follows:

• It is assumed that the initial, discrete data, which is a sequence $\{S_{0,k}\}_{k\in\mathbb{Z}}$, already represents an approximation at a certain scale that is related to the sampling interval. By convention, this scale is fixed at $j = 0$. Thus, the starting point of the algorithm can be viewed as a function f in V_0 defined by $f(x) = \sum_{k\in\mathbb{Z}} S_{0,k}\varphi(x-k)$.

• The analysis proceeds by calculating the projections of f on the approximation spaces and on the spaces of details at the larger scales, that is to say, on the spaces $\{V_j, W_j\}_{j<0}$. From the identity $V_0 = W_0 \oplus W_{-1} \oplus \cdots \oplus W_{-J} \oplus V_{-J}$, one can reconstruct f exactly by starting with its approximation at the scale $-J$ and adding the sequence of details. To be more precise, we write

$$S_{j,k} = 2^{j/2} \langle f, \varphi_{j,k}\rangle, \qquad (1.35)$$

and

$$D_{j,k} = 2^{j/2} \langle f, \psi_{j,k}\rangle. \qquad (1.36)$$

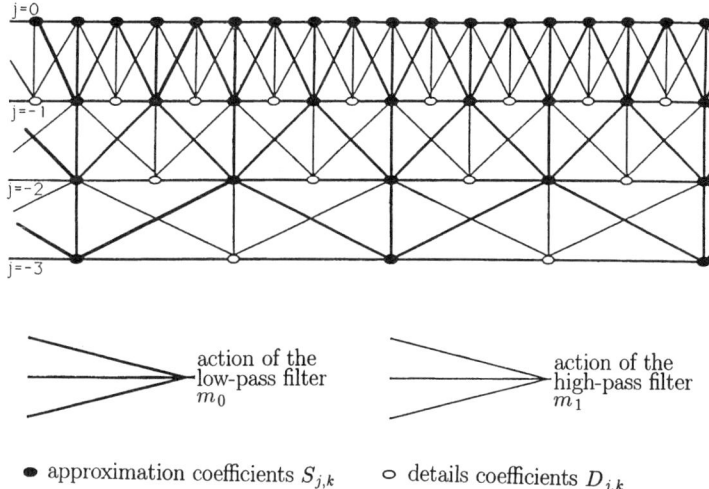

Figure 1.1 *Schematic of the pyramidal algorithm.*

(With this normalization, the approximation coefficients $S_{j,k}$ of a constant C are equal to C for all j.)

We can deduce the following two decomposition formulas from (1.21) and (1.28):

$$S_{j,k} = \sum_{n \in \mathbb{Z}} \overline{h_{n-2k}}\, S_{j+1,n}, \qquad (1.37)$$

and

$$D_{j,k} = \sum_{n \in \mathbb{Z}} \overline{g_{n-2k}}\, S_{j+1,n}. \qquad (1.38)$$

These relations show that the calculation of the coefficients proceeds hierarchically: At each step, from scale $j+1$ to scale j, one applies the low-pass filter $\overline{m_0(\omega)}$ to obtain the approximation and one applies the high-pass filter $\overline{m_1(\omega)}$ to obtain the details; then one applies a decimation operator that keeps only the terms with even index. This last operation eliminates the redundant information in these two sequences. One can then iterate this operation on the approximation that has just been computed and arrive at a new approximation and a new sequence of details.

The pyramidal structure of such an algorithm is illustrated in Figure 1.1. Furthermore, we can see from Figure 1.2 that such a

THE DISCRETE POINT OF VIEW 23

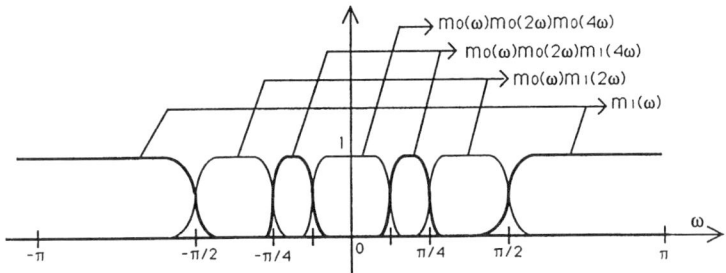

Figure 1.2 *Dyadic division of the spectral band* $[-\pi, \pi]$.

decomposition amounts to dividing the spectrum into dyadic bands by applying the Nyquist rules for decimation as if it were a division into separate bands. In practice this is not the case because the function m_0 is $\mathcal{C}^\infty$ and the channels necessarily overlap. Nevertheless, the global information is conserved, and the exact reconstruction of the original signal is possible. To reconstruct S_{j+1} from the sequences S_j and D_j, we observe that

$$\hat{\varphi}(\omega) = (\overline{m_0(\omega)} + \overline{m_0(\omega+\pi)})\hat{\varphi}(2\omega) + (\overline{m_1(\omega)} + \overline{m_1(\omega+\pi)})\hat{\psi}(2\omega).$$

From this identity we deduce the reconstruction formulas

$$S_{j+1,k} = 2\sum_{n\in\mathbb{Z}} h_{k-2n} S_{j,n} + 2\sum_{n\in\mathbb{Z}} g_{k-2n} D_{j,n}. \tag{1.39}$$

These formulas show that zeros are introduced between the coefficients of the sequences S_j and D_j and the resulting sequences are interpolated with the filters m_0 and m_1.

The set of these operations — filtering, decimation, interpolation, and reconstruction — is often encountered in the signal processing literature under the name 'subband coding scheme.'

REMARKS ABOUT THE FWT ALGORITHM

• In practice, one wishes to use filters with finite impulse response. These are given by the theory of wavelets with compact support that was developed by Ingrid Daubechies (1988). The simplest example is the Haar system.

• To apply this algorithm to a sequence of finite length, one has a choice of several methods:

a) Extend the sequence with zeros. This suffers the defect of creating an artificial discontinuity and requires the computation of coefficients outside the support of the signal to avoid errors at the boundary.

b) Extend the sequence symmetrically about the end points. This will provide economies in the computation to the extent that the filters are also symmetric. Unfortunately, this is not the case for CQFs with finite impulse response. The constructions in Chapter 4 will provide a solution to this problem.

c) Make the sequence and the filters periodic, which corresponds to making the wavelets periodic. This creates *a priori* a discontinuity at the boundaries, but the computation is reduced to computing the essential minimum number of coefficients.

d) Finally, one can use special wavelets that take account of the boundaries. These constructions were developed by Cohen, Daubechies, and Vial (1993), and they allow function spaces on an interval to be characterized by their wavelet coefficients.

• In all of the cases, we see that the pyramidal structure assigns at least half of the computation cost to the initial decomposition step and to the final reconstruction step. Thus the FWT is an algorithm of complexity N, while the FFT has complexity $N \log N$.

• We observe that the algorithm uses the functions φ and ψ only through the Fourier transforms of the filters m_0 and m_1. We are now going to develop a formalism that is tailored to the analysis of the space $l^2(\mathbb{Z})$ and in which the functions φ and ψ do not appear at all.

1.3.2 Multiresolution analyses of $l^2(\mathbb{Z})$

This theory was first introduced by Olivier Rioul. Our objective here is to create, completely analogous to the continuous case, a sequence of approximation spaces for $l^2(\mathbb{Z})$. To maintain the spirit of the algorithm that we have just described, we assume that the initial data corresponds to the resolution $j = 0$. Thus we write $V_0 = l^2(\mathbb{Z})$; the subspaces V_j will be defined by convention for $j \geq 0$, which corresponds here to a decreasing resolution.

The definition of these spaces poses an immediate problem: We wish to define a dilation operator in a way that avoids the explicit introduction of zero values between points and that still preserves the inclusion $V_j \subset V_{j-1}$.

We will now define these operators.

Definition 1.2 *We will call an operator T a dyadic dilation of $l^2(\mathbb{Z})$ if T is an isometry between $l^2(\mathbb{Z})$ and $T(l^2(\mathbb{Z}))$ and if it satisfies the commutation relation*

$$T\tau_k = \tau_{2k}T \quad \text{for all} \quad k \in \mathbb{Z}, \tag{1.40}$$

where τ_k represents translation by k.

We further require that T can be extended to a continuous operator $\tilde{T}$ on $l^\infty(\mathbb{Z})$ and that $\tilde{T}\mathbf{1}_\mathbb{Z} = 1/\sqrt{2}$.

It will become clear that it is sufficient to require that $\tilde{T}\mathbf{1}_\mathbb{Z}$ is a constant. The other conditions will imply that this constant must be $e^{-i\theta}/\sqrt{2}$. For simplicity, we have taken $\theta = 0$.

We will denote the discrete Fourier transform of a sequence $\{s_k\}_{k \in \mathbb{Z}}$ with the corresponding capital letter. Thus we will write $S(\omega) = \sum_{k \in \mathbb{Z}} s_k e^{-ik\omega}$. This spectral representation will allow us to characterize the dyadic dilations.

Proposition 1.2 *Let T be a dyadic dilation and denote the image of the unit impulse at the origin by $T\delta_0 = \{s_k\}_{k \in \mathbb{Z}}$. For any pair of elements $(\{f_k\}, \{g_k\})$ in $(l^2(\mathbb{Z}))^2$,*

$$\{g_k\} = T\{f_k\} \iff G(\omega) = S(\omega)F(2\omega). \tag{1.41}$$

Furthermore, the function $S(\omega)$ is continuous on $\mathbb{R}$ and satisfies

$$|S(\omega)|^2 + |S(\omega + \pi)|^2 = 2, \tag{1.42}$$

and

$$S(0) = \sqrt{2}. \tag{1.43}$$

Proof. Expand $f = \{f_k\}$ in the basis functions $\delta_k \in l^2(\mathbb{Z})$. Because T is continuous (being an isometry) the following relations make sense:

$$\begin{aligned} Tf &= T\left(\sum_{k \in \mathbb{Z}} f_k \delta_k\right) = \sum_{k \in \mathbb{Z}} f_k T\delta_k \\ &= \sum_{k \in \mathbb{Z}} f_k T\tau_k \delta_0 = \sum_{k \in \mathbb{Z}} f_k \tau_{2k} T\delta_0 \\ &= \sum_{k \in \mathbb{Z}} f_k \tau_{2k}\{s_l\} = \sum_{k \in \mathbb{Z}} f_k \{s_{l-2k}\}. \end{aligned}$$

Define $f^* = \{f_k^*\}$ by $f_{2k+1}^* = 0$ and $f_{2k}^* = f_k$. Then

$$\sum_{k \in \mathbb{Z}} f_k \{s_{l-2k}\} = \sum_{k \in \mathbb{Z}} f_k^* \{s_{l-k}\},$$

which means that Tf is the convolution of the sequences f^* and $\{s_k\}$. But $\hat{f}^*(\omega) = F(2\omega)$, and taking the discrete Fourier transform gives (1.41).

From (1.41), the isometry of T is expressed in the spectral domain by

$$\int_0^{2\pi} |F(\omega)|^2 \, d\omega = \int_0^{2\pi} |F(2\omega) S(\omega)|^2 \, d\omega, \qquad (1.44)$$

which by a change of variable becomes

$$2 \int_0^{2\pi} |F(\omega)|^2 \, d\omega = \int_0^{4\pi} |F(\omega)|^2 \, |S(\omega/2)|^2 \, d\omega$$

$$= \int_0^{2\pi} |F(\omega)|^2 \left(|S(\omega/2)|^2 + |S(\omega/2 + \pi)|^2 \right) d\omega.$$

This relation can hold for all 2π-periodic functions F in $L^2[0, 2\pi]$ if and only if we have

$$|S(\omega)|^2 + |S(\omega + \pi)|^2 = 2 \quad \text{for all} \quad \omega \in \mathbb{R}. \qquad (1.45)$$

To show that $S(\omega)$ is continuous, we examine the convergence of $\{s_k\}$ in $l^1(\mathbb{Z})$. Write $s_k = e^{i\theta_k} |s_k|$, and define the sequences $\{e_k\} = \{e^{-i\theta_{-2k}}\}$ and $\{e_k'\} = \{e^{-i\theta_{-(2k+1)}}\}$. (If $s_k = 0$, set the corresponding e_k or e_k' to 0.) Let $\{f_k\} = \tilde{T}\{e_k\}$ and $\{f_k'\} = \tilde{T}\{e_k'\}$. Then $f_0 = \frac{1}{\sqrt{2}} \sum_{k \in \mathbb{Z}} |s_{2k}|$ and $f_{-1}' = \frac{1}{\sqrt{2}} \sum_{k \in \mathbb{Z}} |s_{2k+1}|$. This shows that $\{s_k\}$ converges absolutely and proves that S is continuous. Since $\tilde{T} \mathbf{1}_{\mathbb{Z}} = 1/\sqrt{2}$, we see that

$$\sum_{k \in \mathbb{Z}} s_{2k} = \sum_{k \in \mathbb{Z}} s_{2k+1} = \frac{1}{\sqrt{2}}. \qquad (1.46)$$

This proves that $S(\pi) = 0$ and that $S(0) = \sqrt{2}$. $\square$

It is clear from this argument that the continuity of $\tilde{T}$ is equivalent to $\{s_k\} \in l^1(\mathbb{Z})$. If we had assumed that $\tilde{T} \mathbf{1}_{\mathbb{Z}} = c$ rather than $1/\sqrt{2}$, the same argument shows that $S(\pi) = 0$, and, by (1.45), that $|S(0)| = \sqrt{2}$. Then it follows that $|c| = 1/\sqrt{2}$.

Here we see that the CQFs have reappeared (up to a normalization) as a consequence of the definition of a dyadic dilation. We can now define the multiresolution analyses for $l^2(\mathbb{Z})$.

THE DISCRETE POINT OF VIEW   27

Definition 1.3 *Let T be a dyadic dilation. Write*

$$V_j = T(V_{j-1}) = T^j(l^2(\mathbb{Z})) \quad \text{with} \quad V_0 = l^2(\mathbb{Z}). \qquad (1.47)$$

Since $V_j = T^{j-1}(V_1)$ and $V_1 \subset V_0$, it is clear that $V_j \subset V_{j-1}$. By definition, the sequence $\{V_j\}_{j \in \mathbb{N}}$ is a discrete multiresolution analysis if and only if

$$\bigcap_{j \in \mathbb{N}} V_j = \{0\}. \qquad (1.48)$$

Property (1.48) seems natural, and, indeed, it is the discrete analog of (1.3). Nevertheless, as far as we know, (1.48) is not a trivial consequence of the other assumptions, and it seems to be sensitive to the assumptions about T (or equivalently S). We note, for example, that if we omit the hypothesis $\tilde{T}1_{\mathbb{Z}} = $ a constant, an admissible dilation would be given by $S(\omega) \equiv 1$, which corresponds to the transformation $Tf = g$ with $g_{2k} = f_k$ and $g_{2k+1} = 0$. In this case, $\bigcap_{j \in \mathbb{N}} V_j = \mathbb{C}\,\delta_0 \neq \{0\}$.

To approximate the environment that was established for the continuous theory, we will be concerned with discrete multiresolution analyses that satisfy the following property:

Definition 1.4 *A discrete multiresolution analysis is said to be 'localized' if and only if the function $S(\omega)$ associated with the dilation operator belongs to $\mathcal{C}^\infty(\mathbb{R})$.*

We will see that this condition is sufficient to give us property (1.48).

Note that the regularity of $S(\omega)$ means that the coefficients s_k decrease rapidly, and this is the source of the name 'localized.' On the other hand, if we are given a rapidly decreasing sequence $\{s_k\}$ that satisfies (1.46) and is such that $S(\omega) = \sum_{k \in \mathbb{Z}} s_k e^{-ik\omega}$ satisfies the isometric condition (1.42), then it is easy to see from our arguments that the associated dilation operator T satisfies all of the properties in Definition 1.2.

We are now going to define for discrete signals the analog of the wavelets that were defined for $L^2(\mathbb{R})$.

1.3.3 Orthonormal wavelet bases for $l^2(\mathbb{Z})$

To construct the spaces W_j, which characterize the details at each scale, we will use the function $S(\omega)$ that defines the sequence $\{V_j\}_{j \geq 0}$.

Proposition 1.3 *Let $D(\omega) = e^{-i\omega}\overline{S(\omega + \pi)}$ and let Δ be the operator on $l^2(\mathbb{Z})$ defined by*

$$\{g_k\} = \Delta\{f_k\} \iff G(\omega) = D(\omega)F(2\omega). \tag{1.49}$$

Write $W_j = T^{j-1}\Delta(l^2(\mathbb{Z})) = T^{j-1}(W_1)$. Then W_j is the orthogonal complement of V_j in V_{j-1}.

Proof. It is clear that W_j is included in V_{j-1}. We will first show that the spaces W_j and V_j are orthogonal. But because the operator T is isometric, it is sufficient to show this for V_1 and W_1.

The scalar product of two arbitrary elements v and w of V_1 and W_1 can be calculated by using the discrete Fourier transform as follows:

$$\begin{aligned}\langle v, w \rangle &= \frac{1}{2\pi} \int_0^{2\pi} V(\omega)\overline{W(\omega)}\, d\omega \\ &= \frac{1}{2\pi} \int_0^{2\pi} S(\omega)\overline{D(\omega)}\, F_v(2\omega)\overline{F_w(2\omega)}\, d\omega \\ &= \frac{1}{2\pi} \int_0^{\pi} \left(S(\omega)\overline{D(\omega)} + S(\omega+\pi)\overline{D(\omega+\pi)}\right) F_v(2\omega)\overline{F_w(2\omega)}\, d\omega \\ &= 0.\end{aligned}$$

Similarly, it is sufficient to show that W_1 and V_1 are the complements of each other in $l^2(\mathbb{Z})$. To prove this, we define the adjoint operators T^* and Δ^*. Again, we use the Fourier transform to compute the scalar product $\langle f, Tg \rangle = \langle T^*f, g \rangle$.

$$\begin{aligned}\langle T^*f, g \rangle &= \langle f, Tg \rangle \\ &= \frac{1}{2\pi} \int_0^{2\pi} F(\omega)\overline{S(\omega)G(2\omega)}\, d\omega \\ &= \frac{1}{2\pi} \int_0^{2\pi} \frac{1}{2}\left(F\!\left(\frac{\omega}{2}\right)\overline{S\!\left(\frac{\omega}{2}\right)} + F\!\left(\frac{\omega}{2}+\pi\right)\overline{S\!\left(\frac{\omega}{2}+\pi\right)}\right)\overline{G(\omega)}\, d\omega.\end{aligned}$$

Similarly,

$$\langle \Delta^*f, g \rangle = \frac{1}{2\pi}\int_0^{2\pi} \frac{1}{2}\left(F\!\left(\frac{\omega}{2}\right)\overline{D\!\left(\frac{\omega}{2}\right)} + F\!\left(\frac{\omega}{2}+\pi\right)\overline{D\!\left(\frac{\omega}{2}+\pi\right)}\right)\overline{G(\omega)}\, d\omega.$$

From this we see that these operators correspond to filtering by $\overline{S(\omega)}$ and $\overline{D(\omega)}$ followed by a decimation. In addition, a simple computation shows that

$$TT^* + \Delta\Delta^* = I, \tag{1.50}$$

and
$$T^*T = \Delta^*\Delta = I. \tag{1.51}$$
This means that TT^* and $\Delta\Delta^*$ are the projections of V_0 onto V_1 and W_1 and that the sum of the two projections is the identity. Thus, V_1 and W_1 are the orthogonal complements of each other in V_0, which is what we wished to prove. □

Note that equation (1.50) corresponds to the exact reconstruction formula in the FWT algorithm.

The orthonormal bases for the spaces V_j and W_j can be constructed be starting with the basis $\{\delta_k\}_{k\in\mathbb{Z}}$ for V_0. It suffices to consider
$$\{\alpha_{j,k}\}_{k\in\mathbb{Z}} = \{T^j\delta_k\}_{k\in\mathbb{Z}} = \{\tau_{2^j k}\alpha_{j,0}\}_{k\in\mathbb{Z}} \tag{1.52}$$
for V_j and
$$\{\beta_{j,k}\}_{k\in\mathbb{Z}} = \{T^{j-1}\Delta\delta_k\}_{k\in\mathbb{Z}} = \{\tau_{2^j k}\beta_{j,k}\}_{k\in\mathbb{Z}} \tag{1.53}$$
for W_j.

Here, the integer k represents a translation and refers to an element of $l^2(\mathbb{R})$. Specifically, for each value of k, δ_k, $\alpha_{j,k}$ and $\beta_{j,k}$ represent different elements of $l^2(\mathbb{R})$, and, hence, different sequences.

Note that the discrete Fourier transforms of these elements can be written as
$$A_{j,k}(\omega) = S(\omega)S(2\omega)\cdots S(2^j\omega)\,e^{-i2^j k\omega}, \tag{1.54}$$
$$B_{j,k}(\omega) = S(\omega)S(2\omega)\cdots S(2^{j-1}\omega)D(2^j\omega)\,e^{-i2^j k\omega}. \tag{1.55}$$

This shows us that the decomposition in the FWT algorithm (Figures 1.1 and 1.2) amounts to computing the coefficients in these discrete bases. Projection onto the spaces V_j and W_j corresponds to reconstruction starting from the corresponding channels.

To prove that the family of sequences $\{\beta_{j,k}\}_{j>0, k\in\mathbb{Z}}$ forms an orthonormal basis of discrete wavelets for $l^2(\mathbb{Z})$, it remains for us to show that the intersection of the V_j is the zero element. Then we will have $l^2(\mathbb{Z}) = \overset{\perp}{\oplus}_{j>0} W_j$.

Theorem 1.2 *Assume that the function $S(\omega)$ is in $\mathcal{C}^\infty(\mathbb{R})$. Then*
$$\bigcap_{j\in\mathbb{N}} V_j = \{0\}, \tag{1.56}$$
the corresponding multiresolution analysis is localized, and the discrete wavelets decrease rapidly at infinity.

Proof. First, it is clear from (1.55) that the $B_{j,k}$ are in $\mathcal{C}^\infty(\mathbb{R})$, and, hence, that the wavelets $\{\beta_{j,k}\}_{j>0, k\in\mathbb{Z}}$ decrease rapidly.

For the proof of the main part of the theorem, let P_j be the orthogonal projection of $l^2(\mathbb{Z})$ onto V_j. We wish to show that $P_j f$ tends to 0 when j tends to $+\infty$ for all elements f of $l^2(\mathbb{Z})$. Because the finite sequences are dense in $l^2(\mathbb{Z})$, it is sufficient to show that $P_j \delta_k$ tends to 0 for each fixed k.

We can evaluate the norm of $P_j \delta_k$ by again passing to the spectral domain:

$$\|P_j \delta_k\|^2 = \sum_{l\in\mathbb{Z}} |\langle \delta_k, \alpha_{j,l}\rangle|^2$$

$$= \frac{1}{4\pi^2} \sum_{l\in\mathbb{Z}} \left| \int_0^{2\pi} \overline{S(\omega)} \cdots \overline{S(2^j\omega)}\, e^{i(2^j l - k)\omega}\, d\omega \right|^2.$$

We will use several intermediate results to prove that this sum tends to zero as j tends to $+\infty$ for each fixed k. The first of these reduces the problem to showing that each term in the series tends to 0. □

Proposition 1.4 *For each fixed k, $|\langle \delta_k, \alpha_{j,l}\rangle| \leq \frac{C}{1+|l|}$, where the constant C is independent of j.*

Proof. If $2^j l - k = 0$, we use the Schwarz inequality and the isometry of T to get

$$|\langle \delta_k, \alpha_{j,l}\rangle| \leq \frac{1}{\sqrt{2\pi}} \left(\int_0^{2\pi} |S(\omega)|^2 \cdots |S(2^j\omega)|^2\, d\omega \right)^{1/2} = 1.$$

If $2^j l - k \neq 0$, we can integrate by parts, which gives

$$|\langle \delta_k, \alpha_{j,l}\rangle| = \frac{1}{2\pi|2^j l - k|} \left| \int_0^{2\pi} \frac{d}{d\omega}\left(\prod_{m=0}^{j} S(2^m\omega) \right) e^{i(2^j l - k)\omega}\, d\omega \right|$$

$$\leq \frac{1}{\sqrt{2\pi}|2^j l - k|} \left\| \frac{d}{d\omega}\left(\prod_{m=0}^{j} S(2^m\omega) \right) \right\|_{L^2[0,2\pi]}$$

$$\leq \frac{\sup |\frac{dS}{d\omega}|}{\sqrt{2\pi}|2^j l - k|} \sum_{n=0}^{j} 2^n \left\| \prod_{\substack{m=0 \\ m\neq n}}^{j} S(2^m\omega) \right\|_{L^2[0,2\pi]}.$$

Next, we will show that the norm of the truncated product $\prod_{\substack{m=0 \\ m\neq n}}^{j} S(2^m\omega)$ in $L^2[0,2\pi]$ is $\sqrt{2\pi}$. This is a direct consequence

THE DISCRETE POINT OF VIEW   31

of the isometry of T. To see this, write
$$F(\omega) = S(\omega)S(2\omega)\cdots S(2^{j-n-1}\omega).$$
Then $F(\omega) = \sum f_k e^{-ik\omega}$, where $f = \{f_k\}$ and $f = T^{j-n}\delta_0$. Because T is an isometry, $\sum |f_k|^2 = 1$.

The function G defined by $G(\omega) = F(2\omega)$ is the discrete Fourier transform of $g \in l^2(\mathbb{Z})$ defined by $g_{2k} = f_k$ and $g_{2k+1} = 0$. Clearly, g also has norm 1 in $l^2(\mathbb{Z})$, and, again by the isometry of T, $T^n g$ has norm 1. The truncated product $\prod_{\substack{m=0 \\ m \neq n}}^{j} S(2^m \omega)$ is equal to $S(\omega)\cdots S(2^{n-1}\omega)G(2^n\omega)$, which is the discrete Fourier transform of $T^n g$. Consequently, the norm of the product in $L^2[0, 2\pi]$ is $\sqrt{2\pi}$.

Thanks to this estimate, the scalar product $\langle \delta_k, \alpha_{j,l}\rangle$, is dominated, for fixed k, by
$$|\langle \delta_k, \alpha_{j,l}\rangle| \leq \frac{2^{j+1}\sup\left|\frac{dS}{d\omega}\right|}{|2^j l - k|} \leq \frac{C}{1+|l|}.$$

Thus the series $\sum_{l \in \mathbb{Z}} |\langle \delta_k, \alpha_{j,l}\rangle|^2$ is dominated by a convergent series uniformly in j. By the dominated convergence theorem (applied to the Lebesque space $l^1(\mathbb{Z})$), to prove that the sequence converges to zero, it is sufficient to show that each term $|\langle \delta_k, \alpha_{j,l}\rangle|^2$ tends to zero. We begin by establishing the following result.

Proposition 1.5 *The product $S^j(\omega) = \prod_{m=0}^{j} S(2^m \omega)$ tends to 0 almost everywhere.*

Proof. We observe that the transformation $\omega \mapsto 2\omega$ modulo 2π is ergodic, and this implies that, for almost all $\omega \in [0, 2\pi]$,
$$\begin{aligned}\lim_{j \to +\infty} \frac{\log |S^j(\omega)|}{j} &= \lim_{j \to +\infty} \frac{\sum_{m=0}^{j} \log |S(2^m\omega)|}{j} \\ &= \frac{1}{2\pi}\int_0^{2\pi} \log|S(\omega)|d\omega.\end{aligned} \quad (1.57)$$

Thus the average value of $\log|S(\omega)|$ on $[0, 2\pi]$ comes into play. From the properties of S, we see that
$$\begin{aligned}\frac{1}{2\pi}\int_0^{2\pi} \log(|S(\omega)|)\,d\omega &= \frac{1}{4\pi}\int_0^{2\pi} \log(|S(\omega)|^2)\,d\omega \\ &= \frac{1}{4\pi}\int_0^{\pi}\left(\log(|S(\omega)|^2) + \log(|S(\omega+\pi)|^2)\right)d\omega.\end{aligned}$$

This last integral is equal to $\frac{1}{4\pi}\int_0^\pi \log(|S(\omega)|^2(2-|S(\omega)|^2))\,d\omega$. Since $|S(\omega)|^2(2-|S(\omega)|^2)$ is always less than or equal to 1, we see that the average value of $\log|S(\omega)|$ is less than or equal to 0. It cannot be 0 because that would mean that $|S(\omega)|=1$ almost everywhere, which contradicts Proposition 1.2. Thus,

$$\frac{1}{2\pi}\int_0^{2\pi}\log|S(\omega)|\,d\omega < 0. \tag{1.58}$$

Consequently,
$$\lim_{j\to+\infty}\log|S^j(\omega)| = -\infty, \tag{1.59}$$
and
$$\lim_{j\to+\infty} S^j(\omega) = 0 \tag{1.60}$$
for almost all ω. $\square$

We can now use the following classical lemma to complete the proof of the theorem.

Lemma 1.1 *Let $\{f_n\}$ be a sequence of functions in $L^2[0,2\pi]$ such that:*
- $\|f_n\|_{L^2} = a\ constant,$
- $f_n(\omega)$ *tends to 0 for almost all* ω.

Then f_n converges to 0 in $L^1[0,2\pi]$.

This result can be proved by applying the theorem of Egorov: The sequence f_n tends to 0 uniformly (and hence in L^1) on $[0,2\pi]$ except for a set E of arbitrarily small measure. On E we apply the Schwarz inequality

$$\int_E |f_n(x)|\,dx \leq \|\mathbf{1}_E\|_{L^2}\|f_n\|_{L^2}$$

and conclude that $\|f_n\|_{L^1}$ is arbitrarily small for all sufficiently large n.

Lemma 1.1 applies directly to the sequence $\{S^j(\omega)\}_{j\in\mathbb{N}}$ since $\|S^j(\omega)\|_{L^2} = \sqrt{2\pi}$. From this we deduce that

$$\lim_{j\to+\infty}|\langle\delta_k,\alpha_{j,l}\rangle| = \lim_{j\to+\infty}\int_0^{2\pi}|S^j(\omega)|\,d\omega = 0, \tag{1.61}$$

and this completes the proof of the theorem. $\square$

Before drawing some conclusions from this first chapter, we are going to briefly describe the generalization of these constructions to several variables. This will be used in Chapter 5 and studied in more detail in Appendix B.

1.4 The multivariate case

The axioms that define multiresolution analyses are easily generalized to the function spaces $L^2(\mathbb{R}^n)$ and sequence spaces $l^2(\mathbb{Z}^n)$ of several variables.

In the continuous case, it is clear that properties (1.1) through (1.4) remain unchanged. Property (1.5) becomes

$$\text{There exists a function } g(x) \text{ in } V_0 \text{ such that} \quad (1.62)$$
$$\{g(x-k)\}_{k \in \mathbb{Z}^n} \text{ is a Riesz basis for } V_0.$$

One can then construct a scaling function φ, which satisfies

$$\sum_{l \in \mathbb{Z}^n} |\hat{\varphi}(\omega + 2l\pi)|^2 = 1. \quad (1.63)$$

The transfer function $m_0(\omega) = m_0(\omega_1, \omega_2, \ldots, \omega_n)$ will then satisfy the relation

$$\sum_{(\varepsilon_1,\ldots,\varepsilon_n) \in \{0,1\}^n} |m_0(\omega_1 + \varepsilon_1\pi, \ldots, \omega_n + \varepsilon_n\pi)|^2 = 1. \quad (1.64)$$

Similarly, the function $S(\omega)$, which is used in the construction of the discrete multiresolution analyses, will satisfy the relation

$$\sum_{(\varepsilon_1,\ldots,\varepsilon_n) \in \{0,1\}^n} |S(\omega_1 + \varepsilon_1\pi, \ldots, \omega_n + \varepsilon_n\pi)|^2 = 2^n. \quad (1.65)$$

The difficulty arises when one wishes to construct the wavelets associated with these multiresolution analyses. Since going from one scale to the next involves a loss of information in the ratio of 1 to 2^n, it is necessary to have $2^n - 1$ wavelets (or, equivalently, $2^n - 1$ filters conjugate with $m_0(\omega)$) to carry this lost information. For example, in the case of $l^2(\mathbb{Z}^2)$, an element $f \in V_1$ 'lives on' four times as many points as $P_0 f \in V_0$ (the projection of f onto V_0), and one needs three wavelets to fill in the missing information.

It is easy to verify that the conjugate condition is given as follows: Let $\{\beta_j\}_{j=0,\ldots,2^n-1}$ denote the set of n-tuples $(\varepsilon_1, \ldots, \varepsilon_n)$ with $\varepsilon_k \in \{0,1\}$, and let $\{m_p(\omega)\}_{p=1,\ldots,2^n-1}$ be the desired filters. Then the matrix

$$U(\omega) = (m_i(\omega + \beta_j\pi))_{i,j=0,\ldots,2^n-1} \quad (1.66)$$

must be unitary for all ω.

It is possible to choose the functions $m_p(\omega)$ explicitly for two and three variables (as it is for the univariate case) when $m_0(\omega)$

takes only real values (see Meyer (1990)). A more general solution was obtained by K. Gröchenig (1987), but the conjugate filters are no longer associated with $m_0(\omega)$ by a simple relation. This poses, a priori, a problem for the analysis of $L^2(\mathbb{R}^n)$ and $l^2(\mathbb{Z}^n)$ if one wishes to use filters that have finite impulse response.

This problem does not exist in the special case where the multiresolution analysis is constructed with tensor products. In fact, by using the elements that we have presented for the univariate case, i.e., the spaces V_j and the functions φ, ψ, m_0, m_1, it is possible to define a multiresolution analysis of $L^2(\mathbb{R}^n)$ by

$$\mathcal{V}_j = (\otimes)^n V_j. \tag{1.67}$$

This means that the scaling function and the CQF are then defined by

$$\Phi(x_1, \ldots, x_n) = \varphi(x_1) \cdots \varphi(x_n), \tag{1.68}$$

and

$$M_0(\omega_1, \ldots, \omega_n) = m_0(\omega_1) \cdots m_0(\omega_n). \tag{1.69}$$

The conjugate filters are given by

$$M_\varepsilon(\omega_1, \ldots, \omega_n) = m_{\varepsilon_1}(\omega_1) \cdots m_{\varepsilon_n}(\omega_n), \tag{1.70}$$

where $\varepsilon = (\varepsilon_1, \ldots, \varepsilon_n)$ ranges over $\{0, 1\}^n - (0, \ldots, 0)$. The wavelets ψ_ε are derived from these filters and are multivariate products of the functions φ and ψ.

Such a construction has the advantage that the filtering operations are done separately on each variable. This provides a noticeable improvement in the performance of the algorithms when the filters are long.

Finally, we mention an interesting construction originally due to J. C. Feauveau. He replaces the scaling factor 2 by the dilation matrix

$$\begin{pmatrix} 1 & -1 \\ 1 & 1 \end{pmatrix} = \begin{pmatrix} 0 & 2 \\ -2 & 0 \end{pmatrix}^{1/2} = \begin{pmatrix} 16 & 0 \\ 0 & 16 \end{pmatrix}^{1/8}$$

in the case of two variables. This construction has a decimation rate of 2 and, thus, requires only one wavelet.

1.5 Conclusions

We have presented the discrete theory first by using the continuous point of view (in the section on the FWT) and then by setting it aside completely. We note that this is a theory of digital filtering and that the initial approaches of the engineers who introduced it never took into account the underlying 'functional analytic' aspect of the theory.

We further note that the CQFs that are not associated with multiresolution analyses of $L^2(\mathbb{R})$ — for example the filters given by the construction in Proposition 1.1 with $m_0(\omega) = e^{-i\omega/2} \cos \frac{3\omega}{2}$ — do, on the other hand, provide an analysis of $l^2(\mathbb{Z})$. In fact, all of the conditions for exact reconstruction are satisfied, and, furthermore, these filters satisfy the hypotheses of Theorem 1.2.

Consequently, it seems legitimate to restate questions about the effective contribution of the theory of wavelets to the area of digital signal processing. More precisely, we can formulate several questions that arise naturally:

- Is it possible to characterize exactly those CQFs that are associated with localized multiresolution analyses? Thus the construction of orthonormal wavelets would be reduced to having in hand a function m_0 with 'good properties.'

- Is it possible to classify multiresolution analyses in terms of the properties of m_0? In particular, one would want to know if the wavelet ψ is regular, localized in space or in frequency, has vanishing moments, etc.

- What is the significance, from the point of view of the pyramidal algorithms, of having an underlying multiresolution analysis of $L^2(\mathbb{R})$? At what point do the functions φ and ψ truly intervene in practical applications?

- Finally, can the connections between functional analysis and digital signal processing be extended to a more general setting than that of CQFs and orthonormal wavelets?

The first two questions will be treated in Chapters 2 and 3 respectively. We will take up the third question in Chapters 3 and 5. The last question will be the starting point for Chapter 4.

CHAPTER 2

Wavelets and conjugate quadrature filters

2.1 Introduction

This chapter is devoted to characterizing those conjugate quadrature filters that are associated with localized multiresolution analyses and, hence, with orthonormal wavelets. In view of Theorem 1.1, this amounts to characterizing those filters m_0 that come from a scaling function φ that satisfies the localization property (1.18) and whose integer translates form an orthonormal system.

This problem has two equivalent formulations:
- In the spectral domain, we look for those functions m_0 that are $\mathcal{C}^\infty$ and 2π-periodic and that satisfy

$$|m_0(\omega)|^2 + |m_0(\omega + \pi)|^2 = 1 \tag{2.1}$$

and

$$m_0(0) = 1. \tag{2.2}$$

We also require that if φ is defined by

$$\hat{\varphi}(\omega) = \prod_{k=1}^{+\infty} m_0(2^{-k}\omega) \tag{2.3}$$

then $\hat{\varphi}$ is in all of the Sobolev spaces $H^m(\mathbb{R})$ and

$$\langle \varphi(x), \varphi(x - k) \rangle = \delta_{0,k}. \tag{2.4}$$

- In the spatial domain, we look for those sequences $\{h_n\}_{n \in \mathbb{Z}}$ that decrease rapidly, that satisfy

$$2 \sum_{n \in \mathbb{Z}} h_n \overline{h}_{n+2k} = \delta_{0,k} \tag{2.5}$$

and

$$\sum h_n = 1, \tag{2.6}$$

and for which there exists an L^2 function φ that satisfies (2.4) and the following three properties:

$$\int (1+|x|)^m |\varphi(x)|^2 \, dx < +\infty \quad \text{for all } m, \tag{2.7}$$

$$\hat{\varphi}(0) = \int \varphi(x) \, dx = 1, \tag{2.8}$$

and

$$\varphi(x) = 2 \sum h_n \, \varphi(2x - n). \tag{2.9}$$

It is clear that $\{h_n\}_{n \in \mathbb{Z}}$ is the sequence of Fourier coefficients of the function m_0.

Here we will develop the first approach. This turns out to be quite fruitful because it yields four equivalent results that are very different in nature.

The 'spatial' formulation will be used in the following chapter to study how the regularity of the functions φ and ψ influences the FWT pyramid algorithms. There we will see that the conditions we place on the filters so that they generate orthonormal bases are, in a certain sense, the minimum that is required for most applications.

We begin our presentation with the most general results. We later consider two specific classes of CQFs: the case, which is called the 'finite case,' where the function m_0 vanishes only at isolated points, and the case of wavelets with compact support where the impulse response $\{h_n\}_{n \in \mathbb{Z}}$ is finite.

2.2 The general case

Here we will give a precise characterization of those filters that are associated with localized multiresolution analyses. For this, we need the following definition.

Definition 2.1 *A compact subset K of $\mathbb{R}$ is said to be congruent to $[-\pi, \pi]$ modulo 2π if and only if K is a finite union of intervals and, for almost all ω in $[-\pi, \pi]$, there exists a unique ν in K such that $\omega - \nu \in 2\pi\mathbb{Z}$.*

It is fairly easy to see that such a set K is constructed by partitioning the interval $[-\pi, \pi]$ into a finite number of subintervals and translating each of them by a multiple of 2π, as illustrated in Figure 2.1.

It follows from this that if f is a 2π-periodic function in $L^1[0, 2\pi]$, then

$$\int_K f(\omega) \, d\omega = \int_{-\pi}^{\pi} f(\omega) \, d\omega. \tag{2.10}$$

THE GENERAL CASE

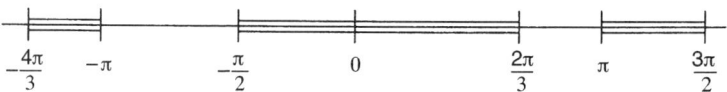

Figure 2.1 *A compact set congruent to $[-\pi,\pi]$ modulo 2π.*

In particular, by taking $f \equiv 1$ we see that the measure of K is 2π. We can now state the main theorem.

Theorem 2.1 *Let m_0 be a 2π-periodic, $\mathcal{C}^\infty$ function that satisfies*

$$|m_0(\omega)|^2 + |m_0(\omega + \pi)|^2 = 1, \tag{2.1}$$
$$m_0(0) = 1, \tag{2.2}$$

and the following property :

> *There exists a compact set K congruent to $[-\pi,\pi]$ modulo 2π whose interior contains 0 and such that, for all j in $\mathbb{N}^*$ and for all ω in K, $m_0(2^{-j}\omega) \neq 0$.* (P)

Under these conditions, the filter generates a localized multiresolution analysis. Conversely, the conjugate quadrature filters that come from multiresolution analyses satisfy the conditions indicated above. Thus, these conditions are necessary and sufficient for a filter m_0 to be associated with a localized multiresolution analysis.

Proof. a) For the first part of the proof, we assume that the CQF m_0 satisfies property (P), and we will show that this filter generates a multiresolution analysis. Since m_0 is a $\mathcal{C}^\infty$ function, the product

$$\hat{\varphi}(\omega) = \prod_{k=1}^{+\infty} m_0(2^{-k}\omega) \tag{2.3}$$

converges uniformly on compact sets to a $\mathcal{C}^\infty$ function that is clearly bounded by 1. This allows us to define φ as a tempered distribution. We next introduce a sequence of functions $\{f_n\}_{n \in \mathbb{N}^*}$ whose terms will be shown to approximate φ as n tends to ∞. We write

$$\hat{f}_n(\omega) = \prod_{k=1}^{n} m_0(2^{-k}\omega)\, \mathbf{1}_{2^n K}, \tag{2.11}$$

where $\mathbf{1}_{2^n K}$ is the characteristic function of the compact set K (from the hypothesis (P)) dilated by the factor 2^n.

We first show that the integer translates of the functions f_n satisfy
$$\langle f_n(x), f_n(x-l) \rangle = \delta_{0,l}. \tag{2.12}$$
For this, we evaluate the expression
$$\int |\hat{f}_n(\omega)|^2 \, e^{il\omega} \, d\omega = 2\pi \, \langle f_n(x), f_n(x-l) \rangle. \tag{2.13}$$

By changing variables appropriately and by using properties (2.1) and (2.10), we have, for $n \geq 2$, the following sequence of relations:

$$\int |\hat{f}_n(\omega)|^2 \, e^{il\omega} \, d\omega$$

$$= \int_{2^n K} \Big(\prod_{k=1}^{n} |m_0(2^{-k}\omega)|^2 \Big) \, e^{il\omega} \, d\omega$$

$$= 2^n \int_{K} \Big(\prod_{k=0}^{n-1} |m_0(2^{k}\omega)|^2 \Big) \, e^{i2^n l\omega} \, d\omega$$

$$= 2^n \int_{-\pi}^{\pi} \Big(\prod_{k=0}^{n-1} |m_0(2^{k}\omega)|^2 \Big) \, e^{i2^n l\omega} \, d\omega$$

$$= 2^n \int_{-\frac{\pi}{2}}^{\frac{\pi}{2}} \Big(\prod_{k=1}^{n-1} |m_0(2^{k}\omega)|^2 \Big) \big(|m_0(\omega)|^2 + |m_0(\omega+\pi)|^2 \big) e^{i2^n l\omega} \, d\omega$$

$$= 2^{n-1} \int_{-\pi}^{\pi} \Big(\prod_{k=0}^{n-2} |m_0(2^{k}\omega)|^2 \Big) \, e^{i2^{n-1} l\omega} \, d\omega$$

$$= \int |\hat{f}_{n-1}(\omega)|^2 \, e^{il\omega} \, d\omega.$$

For $n = 1$, the same manipulations show that
$$\int |\hat{f}_1(\omega)|^2 \, e^{il\omega} \, d\omega = \int_{-\pi}^{\pi} e^{il\omega} \, d\omega = 2\pi \, \delta_{0,l}.$$

Consequently, by induction, the f_n satisfy property (2.4). We next show that this holds for the function φ.

Note that, since the origin is an interior point of K, the set $2^n K$ eventually covers every point of $\mathbb{R}$ as n tends to $+\infty$. It follows from this that $\hat{f}_n(\omega)$ converges to $\hat{\varphi}(\omega)$ for each fixed ω. We can now apply Fatou's lemma: Since the L^2 norms of the functions f_n

THE GENERAL CASE

are all equal to 1, $\hat{\varphi}$ and φ are also square integrable and

$$\int |\hat{\varphi}(\omega)|^2 \, d\omega \leq 2\pi. \tag{2.14}$$

Property (P) allows us to be more explicit about the convergence of the f_n to φ. We note that this property is equivalent, using (2.3), to saying that $|\hat{\varphi}(\omega)|$ is bounded away from zero on K. Thus, there is a $C > 0$ such that

$$|\hat{\varphi}(\omega)| > C > 0 \quad \text{for all } \omega \text{ in } K. \tag{2.15}$$

By observing that

- $\omega \notin 2^n K \Rightarrow \hat{f}_n(\omega) = 0$ and

- $\omega \in 2^n K \Rightarrow \hat{\varphi}(\omega) = \hat{f}_n(\omega) \, \hat{\varphi}(2^{-n}\omega)$,

we establish that

$$|\hat{f}_n(\omega)| \leq \frac{|\hat{\varphi}(\omega)|}{C} \quad \text{for all } \omega. \tag{2.16}$$

We can now apply Lebesque's dominated convergence theorem and conclude that f_n converges to φ in $L^2(\mathbb{R})$ and, as a consequence, that the function φ and its translates form an orthonormal system.

It remains to show that $\hat{\varphi}$ is in $H^m(\mathbb{R})$, for all m in $\mathbb{N}$. For this, we will again use a sequence of functions that converge to φ, but in this case the truncations will be much more regular than the characteristic functions that were used before. This will allow us to differentiate the approximating functions. We choose a function $\theta(\omega)$ having the following properties:

- $\theta(\omega)$ is C^∞ and $0 \leq \theta(\omega) \leq 1$
- $\theta(\omega) = 1$ if $\omega \in [-\pi, \pi]$
- $\theta(\omega) = 0$ if $\omega \notin [-2\pi, 2\pi]$

and we define the approximating sequence by

$$u_n(\omega) = \theta(2^{-n}\omega) \prod_{k=1}^{n} m_0(2^{-k}\omega). \tag{2.17}$$

It is clear from this definition that all of the successive derivatives of the u_n tend pointwise to those of $\hat{\varphi}$.

We first show that the L^2 norms of the $\frac{du_n}{d\omega}$ are bounded uniformly in n. Differentiating the product (2.17) and taking the obvious estimates gives

$$\left\|\frac{du_n}{d\omega}\right\| \leq 2^{-n}\left\|\frac{d\theta}{d\omega}(2^{-n}\omega)\prod_{k=1}^{n}m_0(2^{-k}\omega)\right\|$$

$$+\sum_{l=1}^{n}2^{-l}\left\|\frac{dm_0}{d\omega}(2^{-l}\omega)\theta(2^{-n}\omega)\prod_{\substack{k=1\\k\neq l}}^{n}m_0(2^{-k}\omega)\right\|$$

$$\leq 2^{-n}\sup\left|\frac{d\theta}{d\omega}\right|\left[2\int_{-2^n\pi}^{2^n\pi}\prod_{k=1}^{n}|m_0(2^{-k}\omega)|^2\,d\omega\right]^{1/2}$$

$$+\sum_{l=1}^{n}2^{-l}\sup\left|\frac{dm_0}{d\omega}\right|\left[2\int_{-2^n\pi}^{2^n\pi}\prod_{\substack{k=1\\k\neq l}}^{n}|m_0(2^{-k}\omega)|^2\,d\omega\right]^{1/2}$$

Next, we must evaluate the norms in $L^2[-2^n\pi, 2^n\pi]$ of the products $\prod_{\substack{k=1\\k\neq l}}^{n}m_0(2^{-k}\omega)$. Using the same techniques as were used in the evaluation of $\int|\hat{f}_n(\omega)|^2 e^{il\omega}\,d\omega$, we find that

$$\int_{-2^n\pi}^{2^n\pi}\prod_{\substack{k=1\\k\neq l}}^{n}|m_0(2^{-k}\omega)|^2\,d\omega = 2^n\int_{-\pi}^{\pi}\prod_{\substack{k=0\\k\neq n-l}}^{n-1}|m_0(2^k\omega)|^2\,d\omega$$

$$= 2^{n-1}\int_{-\pi}^{\pi}\prod_{\substack{k=0\\k\neq n-l-1}}^{n-2}|m_0(2^k\omega)|^2\,d\omega$$

$$= \cdots = 2^l\int_{-\pi}^{\pi}\prod_{k=1}^{l-1}|m_0(2^k\omega)|^2\,d\omega$$

$$= 2^{l-1}\int_{-2\pi}^{2\pi}\prod_{k=0}^{l-2}|m_0(2^k\omega)|^2\,d\omega$$

$$= 2^l\int_{-\pi}^{\pi}\prod_{k=0}^{l-2}|m_0(2^k\omega)|^2\,d\omega$$

$$= \cdots = 4\pi.$$

These computations show that removing one of the factors increases the L^2 norm of the product by the factor $\sqrt{2}$ and that these products are bounded in L^2 independently of their length. Combined with the previous inequality, this allows us to bound

$\left\|\frac{du_n}{d\omega}\right\|$ uniformly in n. In fact,

$$\left\|\frac{du_n}{d\omega}\right\| \leq 2^{-n} \sup\left|\frac{d\theta}{d\omega}\right|(4\pi)^{1/2} + \sup\left|\frac{dm_0}{d\omega}\right|(8\pi)^{1/2} \sum_{l=1}^{n} 2^{-l}.$$

Once again, applying Fatou's lemma shows that $\frac{d\hat{\varphi}}{d\omega}$ belongs to $L^2(\mathbb{R})$.

The generalization of this result to the higher-order derivatives poses mechanical, but no conceptual, problems. The terms in the expansion of $\left(\frac{d}{d\omega}\right)^m u_n$ have the same form (and are estimated the same way) as those for the expansion of $\left(\frac{d}{d\omega}\right)^{m-1} u_n$, except the term that contains the derivative of the product

$$\prod_{\substack{k=1 \\ k \neq l_1, \ldots, l_{m-1}}}^{n} m_0(2^{-k}\omega).$$

Differentiating this term introduces another geometric series, and one is led to evaluate the $L^2[-2^n\pi, 2^n\pi]$ norms of the products $\prod_{k=1(k \neq l_1, \ldots, l_m)}^{n} m_0(2^{-k}\omega)$, which have one more missing factor. From the computations we have already done, it is fairly easy to see that

$$\int_{-2^n\pi}^{2^n\pi} \prod_{\substack{k=1 \\ k \neq l_1, \ldots, l_m}}^{n} |m_0(2^{-k}\omega)|^2 \, d\omega = 2^{m+1}\pi.$$

Although it is quite tedious to write out a full proof by induction, the result is the same: The L^2 norms of the derivatives $\left(\frac{d}{d\omega}\right)^m u_n$ are bounded independently of n, and $\left(\frac{d}{d\omega}\right)^m \hat{\varphi}$ belongs to $L^2(\mathbb{R})$. Thus, $\hat{\varphi}$ belongs to all of the Sobolev spaces $H^m(\mathbb{R})$. This completes the proof of the first part of the theorem.

b) To prove the converse, we assume that we have a localized multiresolution analysis and that m_0 is the associated filter. We must show that m_0 satisfies property (P).

The first step is to show that

$$\sum_{l \in \mathbb{Z}} |\hat{\varphi}(\omega + 2l\pi)|^2 = 1 \tag{2.18}$$

for all ω in $[-\pi, \pi]$. Recall that our hypothesis includes the assumptions that $\hat{\varphi}$ is in all of the Sobolev spaces $H^m(\mathbb{R})$, in particular, $H^1(\mathbb{R})$, and that φ satisfies (2.4). This latter condition implies immediately that (2.18) holds almost everywhere. (This can be seen from the inverse of the argument that goes from (1.14) to (1.16), or, in more detail, from Lemma 4.1.)

In fact, (2.18) is valid for all ω in $[-\pi, \pi]$. To see this, we note that since $m_0(\pi) = 0$, we have $\hat{\varphi}(2\pi n) = 0$ for all $n \in \mathbb{Z} \setminus \{0\}$. Thus, for all ω in $[-\pi, \pi]$ and $n \in \mathbb{Z} \setminus \{0\}$, we have the estimate

$$|\hat{\varphi}(\omega + 2n\pi)| \leq \int_{(2n-1)\pi}^{(2n+1)\pi} |(\hat{\varphi})'(\omega)|\, d\omega$$

and thus, by the Schwarz inequality,

$$|\hat{\varphi}(\omega + 2n\pi)|^2 \leq 2\pi \int_{(2n-1)\pi}^{(2n+1)\pi} |(\hat{\varphi})'(\omega)|^2\, d\omega\,.$$

Since $\hat{\varphi}$ is in H^1, we conclude that the series $\sum_n |\hat{\varphi}(\omega + 2n\pi)|^2$ converges uniformly in $[-\pi, \pi]$ so that it is equal to 1 for all ω.

Consequently, for each ω in $[-\pi, \pi]$, there exists an integer l_ω such that $\hat{\varphi}(\omega + 2l_\omega \pi) \neq 0$. Note that necessarily $l_0 = 0$. Since $\hat{\varphi}$ is continuous, we can also find open intervals V_ω about each point ω such that

$$\mu \in V_\omega \implies |\hat{\varphi}(\mu + 2l_\omega \pi)|^2 \geq C_\omega > 0\,. \tag{2.19}$$

Choose a finite family $\{\omega_0, \ldots, \omega_n\}$, with $\omega_0 = 0$, such that the union of the neighborhoods V_{ω_j} covers $[-\pi, \pi]$. We can now construct a partition of $[-\pi, \pi]$ as follows:
- Let $R_0 = V_0 \cap [-\pi, \pi]$.
- Define recursively $R_j = V_j \cap [-\pi, \pi] \setminus \left(\bigcup_{k=0}^{j-1} R_k\right)$

for $j = 1, \ldots, n$.

If we define the compact set K by

$$K = \overline{\bigcup_{j=0}^{n} \{\omega + 2l_{\omega_j}\pi \mid \omega \in R_j\}}\,, \tag{2.20}$$

it is clear that it is congruent to $[-\pi, \pi]$ modulo 2π, contains a neighborhood of the origin (R_0), and serves to satisfy property (P). In fact,

$$|m_0(2^{-n}\omega)| \geq |\hat{\varphi}(\omega)| > 0 \tag{2.21}$$

for all ω in K and all n in $\mathbb{N}^*$. This completes the proof of Theorem 2.1. □

REMARKS
- This result generalizes easily to several variables. Some applications of this are presented in Appendix B.

• It might seem that property (P) is difficult to verify since it concerns the values of m_0 at all of the points $2^{-n}\omega$ for $n \in \mathbb{N}^*$ and $\omega \in K$. We note that, in fact, since K is compact and $m_0(0) = 1$, one must check that $m_0(2^{-n}\omega) \neq 0$ for only a finite number of n.

• The construction of the compact set K is based essentially on the inequality

$$\sum_{l \in \mathbb{Z}} |\hat{\varphi}(\omega + 2l\pi)|^2 > 0.$$

Consequently, property (P) is satisfied more generally by filters that generate Riesz bases for localized multiresolution analyses.

• If we take the simplest case where $K = [-\pi, \pi]$ and ask that (P) be satisfied, we are led to the condition that m_0 cannot vanish on $[-\pi/2, \pi/2]$. Thus, this condition is sufficient for the construction of a localized multiresolution analysis. This condition, which is not necessary but which is often satisfied in practice, was established by Stéphane Mallat (1989). Theorem 2.1 allows us to generalize this condition as indicated in the following corollary.

Corollary 2.1 *A sufficient condition on the function m_0 so that it generates a localized multiresolution analysis is that*

$$m_0(\omega) \neq 0 \quad \text{if} \quad \omega \in \left[-\frac{\pi}{3}, \frac{\pi}{3}\right]. \tag{2.22}$$

Proof. Let A be the set of zeros of m_0 in $\left[-\frac{\pi}{2}, -\frac{\pi}{3}\right[$ and let B be the set of zeros in $\left]\frac{\pi}{3}, \frac{\pi}{2}\right]$.

For each zero α in A, we can find an open interval V_α containing α such that $|m_0(\omega)| < \frac{\sqrt{2}}{2}$ on V_α. Hence, by (2.1), $|m_0(\omega+\pi)| > \frac{\sqrt{2}}{2}$ on V_α. Since $A \cap \left[-\frac{\pi}{2}, -\frac{\pi}{3}\right]$ is compact, the covering $\bigcup_{\alpha \in A} V_\alpha$ can be reduced to a finite covering. By using the same method that we used in the proof of Theorem 2.1, we can construct a finite partition of disjoint intervals A_j in $\left[-\frac{\pi}{2}, -\frac{\pi}{3}\right]$ that covers A and such that the A_j are contained in $\bigcup_{\alpha \in A} V_\alpha$. Similarly, we can construct a finite set of disjoint intervals B_j contained in $\left[\frac{\pi}{3}, \frac{\pi}{2}\right]$ such that $|m_0(\omega - \pi)| > \frac{\sqrt{2}}{2}$ if $\omega \in B_j$.

We can now define the compact set K that will serve to demonstrate property (P). Let

$$R_a = \{2\omega \mid \omega \in A_j\}, \qquad R_b = \{2\omega \mid \omega \in B_j\},$$
$$S_a = \{2\omega + 2\pi \mid \omega \in A_j\}, \quad S_b = \{2\omega - 2\pi \mid \omega \in B_j\},$$

and define K by

$$K = \overline{\left([-\pi, \pi] \setminus (R_a \cup R_b)\right) \cup (S_a \cup S_b)}. \tag{2.23}$$

Clearly, K is congruent to $[-\pi, \pi]$ modulo 2π and contains a neighborhood of the origin, namely $\left[-\frac{\pi}{3}, \frac{\pi}{3}\right]$. By construction, $m_0(\omega/2)$ does not vanish on K. Furthermore, K is in the interval $\left[-\frac{4\pi}{3}, \frac{4\pi}{3}\right]$ and the other factors $m_0(2^{-k}\omega)$ $(k \geq 2)$ do not vanish on K because of the hypothesis (2.22) regarding m_0.

Property (P) is thus satisfied. Hence, filters that do not vanish on $\left[-\frac{\pi}{3}, \frac{\pi}{3}\right]$ generate multiresolution analyses. □

REMARK

This new sufficient condition represents a limit in the following sense: If the function m_0 vanishes at the points $-\frac{\pi}{3}$ and $\frac{\pi}{3}$, property (P) is no longer satisfied. This includes the counterexample given by the dilated Haar system where $m_0(\omega) = \frac{1+e^{-3i\omega}}{2}$. More specifically, Proposition 1.1 showed us that if m_0 vanished at any of the points $\frac{(2n+1)\pi}{2p+1}$, then it could not generate a multiresolution analysis.

We are now going to investigate the arithmetic aspects of property (P) by considering a very particular class of conjugate quadrature filters.

2.3 The finite case

We assume in this section that the function m_0 vanishes only at isolated points on the real line. We call this the 'finite case' because the hypothesis implies that m_0 has only a finite number of zeros in the interval $[-\pi, \pi]$.

This is the case for filters that have finite impulse response and, more generally, for recursive filters, which are filters whose transfer function is the quotient of two trigonometric polynomials.

We will use this hypothesis to express property (P) arithmetically. We begin with the negation of the property to develop information about the positions of the points where m_0 vanishes.

More precisely, we will see that the transfer functions of filters that do not generate multiresolution analyses assume values on the unit circle $|m_0(\omega)| = 1$ on a non-trivial cyclic orbit of the mapping $\omega \mapsto 2\omega$ modulo 2π. Several preparatory lemmas head us toward this result.

Lemma 2.1 *Let m_0 be a regular 2π-periodic function satisfying the properties (2.1) and (2.2) that characterize the CQFs.*

Let φ be the scaling function defined by (2.3). Then m_0 does not satisfy property (P) if and only if there exists an ω_0 in $[-\pi, \pi]$ such

that
$$\sum_{l \in \mathbb{Z}} |\hat{\varphi}(\omega_0 + 2l\pi)|^2 = 0. \tag{2.24}$$

Proof. The lemma is a direct result of the proof of Theorem 2.1. In fact, the construction of the compact set K is possible if and only if
$$\sum_{l \in \mathbb{Z}} |\hat{\varphi}(\omega_0 + 2l\pi)|^2 \neq 0.$$
□

Since $\hat{\varphi}$ is obtained from the infinite product (2.3), we can use the identity (2.24) to construct a set of points where m_0 vanishes. We begin by noting that for each fixed $l \in \mathbb{Z}$, there exists at least one j greater than or equal to 1 such that $m_0(2^{-j}(\omega_0 + 2l\pi)) = 0$. We can then define
$$j_l = \inf \{ j \geq 1 \mid m_0(2^{-j}(\omega_0 + 2l\pi)) = 0 \}, \tag{2.25}$$
and
$$\nu_l = 2^{-j_l}(\omega_0 + 2l\pi) \qquad \text{modulo } 2\pi. \tag{2.26}$$
The integer j_l corresponds to the first factor in (2.3) that causes $\hat{\varphi}(\omega_0 + 2l\pi)$ to vanish, and ν_l is the associated root of m_0 in the interval $[-\pi, \pi[$. Thus
$$\nu_l = 2^{-j_l}(\omega_0 + 2l\pi) + 2N_l\pi, \tag{2.27}$$
where j_l and N_l are functions of l.

Thus we have a family $\{\nu_j\}_{l \in \mathbb{Z}}$, which, because of the original hypothesis, contains only a finite number of distinct elements. On the other hand, the family $\{j_l\}_{l \in \mathbb{Z}}$ contains an infinite number of elements, as shown by the next result.

Lemma 2.2 *There exists a sequence of integers $\{l_k\}_{k \geq 0}$ such that $j_{l_{k+1}} > j_{l_k}$.*

Proof. We construct the sequence l_k explicitly. Take $l_0 = l$, where l is any element of $\mathbb{Z}$. For ease of notation we write
$$n_k = j_{l_k}, \quad N_k = N_{l_k}, \tag{2.28}$$
and
$$\mu_k = \nu_{l_k} = 2^{-n_k}(\omega_0 + 2l_k\pi) + 2N_k\pi. \tag{2.29}$$

Proceeding by induction, we assume that l_k, n_k, N_k, and μ_k have been determined based on the definition
$$l_{k+1} = l_k + \varepsilon 2^{n_k - 1} + 2N_k\pi, \tag{2.30}$$

where $\varepsilon = 1$. (This parameter will be useful later.) Then
$$2^{n_k}(\mu_k + \varepsilon\pi) = \omega_0 + 2l_{k+1}\pi, \qquad (2.31)$$
and
$$m_0(2^{n_k-n}(\mu_k + \varepsilon\pi)) = m_0(2^{-n}(\omega_0 + 2l_{k+1}\pi)).$$
The first n for which the right side is zero is by definition n_{k+1}, corresponding to l_{k+1}.

We will show that n_{k+1} is strictly larger than n_k. Indeed, for all $n < n_k$,
$$m_0(2^{-n}(\omega_0 + 2l_{k+1}\pi)) = m_0(2^{-n}(\omega_0 + 2l_k\pi) + 2^{n_k-n}(\varepsilon + 2N_k)\pi)$$
$$= m_0(2^{-n}(\omega_0 + 2l_k\pi)) \neq 0,$$
by the definition of n_k. For $n = n_k$,
$$m_0(2^{-n_k}(\omega_0 + 2l_{k+1}\pi)) = m_0(\mu_k + \pi) \neq 0$$
because $m_0(\mu_k) = 0$ and $|m_0(\omega)|^2 + |m_0(\omega+\pi)|^2 = 1$. Consequently, $n_{k+1} > n_k$, and we have an infinite sequence of exponents $\{n_k\}$.

This proof shows that $2^{n_k-n}(\mu_k + \varepsilon\pi)$ falls on one of the ν_l when $n = n_{k+1}$ and that n_{k+1} is the first value for which this is true. Since $n_{k+1} > n_k$, this root, $2^{n_k - n_{k+1}}(\mu_k + \varepsilon\pi)$, is in $[0, \pi[$ for $k \geq 1$. The right-hand side of the relation
$$2^{n_k - n_{k+1}}(\mu_k + \varepsilon\pi) = 2^{-n_{k+1}}(\omega_0 + 2l_{k+1}\pi) \qquad (2.32)$$
is by definition $\mu_{k+1} - 2N_{k+1}\pi$. However, since $2^{-n_{k+1}}(\omega_0 + 2l_{k+1}\pi)$ lies in the interval $[-\pi, \pi[$, $N_{k+1} = 0$, and
$$\mu_{k+1} = 2^{n_k - n_{k+1}}(\mu_k + \varepsilon\pi). \qquad (2.33)$$
This means, in fact, that $N_k = 0$ for all $k \geq 1$ and that (2.33) is true for all $k \geq 1$. Finally, observe that this proof goes through with $\varepsilon = -1$, in which case $\mu_k \in [-\pi, 0[$ for $k \geq 1$.

We will see further on that the sequence $\{j_l\}_{l \in \mathbb{Z}}$ contains all the elements of $\mathbb{N}^*$. For the moment, we use the families $\{n_k\}_{k \geq 0}$ and $\{\mu_k\}_{k \geq 0}$ to determine the arithmetic nature of the roots $\{\nu_l\}_{l \in \mathbb{Z}}$.
□

Lemma 2.3 *The roots ν_l are all of the form $\nu_l = r_l\pi$ where r_l is a rational number.*

Proof. It is sufficient to show that this is true for ω_0. We now know that the family $\{j_l\}_{l \in \mathbb{Z}}$ is infinite. Thus we can find two integers l and l' such that $\nu_l = \nu_{l'}$ and $j_l \neq j_{l'}$. But,
$$2^{-j_l}(\omega_0 + 2l\pi) - 2^{-j_{l'}}(\omega_0 + 2l'\pi) \in 2\mathbb{Z}\pi, \qquad (2.34)$$

THE FINITE CASE 49

which implies that
$$(2^{-j_l} - 2^{-j_{l'}})\omega_0 \in \mathbb{Q}\pi. \qquad (2.35)$$

Hence
$$\omega_0 \in \mathbb{Q}\pi. \qquad (2.36)$$

The numbers $r_l = \frac{1}{\pi}\nu_l$ are thus rationals that lie in the interval $[-1, 1[$. We next show that they have a more specific form. □

Lemma 2.4 *Let $\frac{p_l}{q_l}$ be the irreducible representation of the rational r_l. Then*
- *p_l is odd, and*
- *q_l is odd and strictly greater than 1.*

Proof. We proceed by eliminating the other possible cases using indirect arguments. If r_l were an integer, then ν_l would equal 0 or $-\pi$. Since $\nu_l = 2^{-j_l}(\omega_0 + 2l\pi) + 2N_l\pi$, this would imply that $\omega_0 \in 2\pi\mathbb{Z}$. But this is impossible because

$$\sum_{l \in \mathbb{Z}} |\hat{\varphi}(2l\pi)|^2 = |\hat{\varphi}(0)|^2 = 1. \qquad (2.37)$$

Next suppose that r_l has an even denominator. Construct the sequence $\{\mu_k\}_{k \geq 0}$ following the procedure of Lemma 2.2 starting with $\mu_0 = \nu_l$. The irreducible form of r_l can be written as $2^{-j_0}\frac{p_0}{q_0}$ where p_0 and q_0 are odd and j_0 is greater than or equal to 1. Then $\mu_0 = 2^{-j_0}\frac{p_0}{q_0}\pi$ and, by (2.33),

$$\mu_1 = 2^{n_0-n_1}(\mu_0 + \pi) = 2^{n_0-n_1-j_0}\left(\frac{p_0 + 2^{j_0}q_0}{q_0}\right)\pi$$

$$= 2^{-j_1}\frac{p_1}{q_0}\pi.$$

This gives us the irreducible form of $\frac{\mu_1}{\pi}$ because $p_1 = p_0 + 2^{j_0}q_0$ is odd and relatively prime to q_0.

From Lemma 2.2 it is clear that we have a strict inequality with $j_0 < j_1$. By iterating this process, we see that it is possible to write μ_k as

$$\mu_k = 2^{-j_k}\frac{p_k}{q_0}\pi, \qquad (2.38)$$

where p_k and q_0 are odd and relatively prime and where j_k is an increasing sequence. Consequently, we have generated an infinite collection of different roots of m_0 in $[-\pi, \pi[$. This is a contradiction since there are only a finite number of distinct elements in the family $\{\mu_k\}_{k \geq 0}$.

Finally, suppose that the numerator of r_l is even and proceed with the same construction. Writing $\mu_0 = \frac{p}{q}\pi$, we have

$$\mu_1 = 2^{n_0-n_1}\left(\frac{p+q}{q}\right)\pi. \tag{2.39}$$

Since $p+q$ is odd, we are back in the previous situation and are led to the same contradiction. □

We are now ready to tackle the complete characterization of those roots of m_0 that are the elements of the set $\{\nu_l\}_{l\in\mathbb{Z}}$. The following lemma assures us that a given number j_l corresponds to a single possible zero.

Lemma 2.5 *If $j_l = j_{l'}$ then $\nu_l = \nu_{l'}$.*

Proof. Assume that $j_l = j_{l'}$. Definition (2.26) implies that

$$\nu_l - \nu_{l'} = 2k\pi + 2^{1-j_l}(l-l')\pi. \tag{2.40}$$

This shows that the denominator of $\frac{1}{\pi}(\nu_l - \nu_{l'})$ is either even or 1. But Lemma 2.4 tells us that this same denominator must be odd. Thus it must be 1, which implies that either the numerator is zero or $\nu_l = \nu_{l'} \pm \pi$. But, by (2.1), the last relation contradicts $m_0(\nu_l) = m_0(\nu_{l'}) = 0$. Hence the numerator is zero and $\nu_l = \nu_{l'}$. □

This lemma implies that we can find all possible values of ν_l by finding one value for each element of the family $\{j_l\}_{l\in\mathbb{Z}}$. In particular, if $\{j_l\}_{l\in\mathbb{Z}}$ contains all of the positive integers $k \geq 1$, then it is sufficient to find a root ν_l for each k. We begin by showing that the family $\{j_l\}_{l\in\mathbb{Z}}$ contains the integer 1.

Observe that $\nu_{l+1} - 2^{j_l-j_{l+1}}\nu_l + 2^{-j_{l+1}+1}\pi$ is in $2\pi\mathbb{Z}$ by the definition of ν_l (2.26). Assuming that j_l is greater than or equal to j_{l+1} (interchanging ν_l and ν_{l+1} if necessary), we see that this is possible only if $j_{l+1} = 1$, since by Lemma 2.4 $\frac{1}{\pi}\left(\nu_{l+1} - 2^{j_l-j_{l+1}}\nu_l\right)$ has an odd denominator. Said another way, this means that, depending on the choice of ω_0, $j_l = 1$ for all even l or $j_l = 1$ for all odd l. This argument can be continued to show that $\{j_l\}$ contains all of the positive integers. For example, if $j_l = 1$ for all even l, then one shows that $j_l = 2$ for every other odd l, and so on. This is not necessary for the task at hand, and the fact that $\{j_l\}$ contains all of the positive integers is a byproduct of the next section.

We are now going to construct sequentially all of the roots ν_l. Write $\gamma_1 = \nu_a = \frac{p}{q}\pi$ where $j_a = 1$. The integers p and q are

THE FINITE CASE 51

odd and relatively prime. Next we examine $\gamma_1 + \pi = \frac{p+q}{q}\pi$ and $\gamma_1 - \pi = \frac{p-q}{q}\pi$. It is clear that both $p+q$ and $p-q$ are even, but since their difference is $2q$, one is divisible by 2 and not by 4.

The argument given in Lemma 2.2 for the construction of the sequence $\{\mu_k\}_{k\geq 0}$ shows that we recover an element of the family $\{\nu_l\}_{l\in\mathbb{Z}}$ by dividing $\gamma_1 + \pi$ or $\gamma_1 - \pi$ by 2, 4, 8,... until we 'fall on' a root. But from Lemma 2.4 we know that we arrive at this root exactly when we reach the irreducible representation imposed by Lemma 2.4. For the case at hand, this means that $2^{-1}(\gamma_1 + \pi)$ is a root if $2^{-1}(p+q)$ is odd and that $2^{-1}(\gamma_1 - \pi)$ is a root if $2^{-1}(p-q)$ is odd.

Based on these remarks, we define

- $\gamma_2 = \dfrac{\gamma_1 + \pi}{2}$ if $\dfrac{p+q}{2}$ is odd

- $\gamma_2 = \dfrac{\gamma_1 - \pi}{2}$ if $\dfrac{p-q}{2}$ is odd

The sequence $\{\gamma_k\}_{k\geq 1}$ is defined by induction: $\gamma_k = \frac{p_k}{q}\pi$, where $p_1 = p$ and $\{p_{k+1}\} = \{\frac{p_k+q}{2}, \frac{p_k-q}{2}\} \cap 2\mathbb{Z}+1$.

Mixing the current notation with the notation of Lemma 2.2, we see that

$$\gamma_{k+1} = 2^{n_k - n_{k+1}}(\gamma_k + \varepsilon_k \pi) = 2^{-n_{k+1}}(\omega_0 + 2l_{k+1}\pi)$$

where $l_1 = a$, $l_{k+1} = l_k + \varepsilon_k 2^{n_k-1} + 2^{n_k} N_k$, and ε_k equals -1 or $+1$ according to the rule stated above. With this notation, we have just shown that $n_k - n_{k+1} = -1$. Thus, since $n_1 = 1$, $n_k = k$, and this proves that $\{j_l\}$ contains all of the positive integers. Furthermore, these remarks combined with Lemma 2.5 prove the following:

$$\gamma_k = \nu_l \iff j_l = k. \qquad (2.41)$$

Thus the sequence γ_k takes on all possible values of the ν_l. Since these elements are of the form $\frac{p_k}{q}\pi$ where $-q < p_k < q$, they form a cycle under the transformation $\{p_{k+1}\} = \{\frac{p_k+q}{2}, \frac{p_k-q}{2}\} \cap 2\mathbb{Z}+1$.

This proves that the vanishing of m_0 on a cycle $\{\gamma_1,\ldots,\gamma_n\}$ is a necessary condition that property (P) is not verified. The next theorem will show that this condition is also sufficient. But before stating and proving the theorem, we give some examples of these cycles:

- $\{-\frac{\pi}{3}, \frac{\pi}{3}\}$ and $\{\frac{\pi}{5}, \frac{3\pi}{5}, -\frac{\pi}{5}, -\frac{3\pi}{5}\}$ correspond to the counterexamples in Chapter 1 (Proposition 1.1).

- When the denominator is increased, we get $\{\frac{\pi}{7}, -\frac{3\pi}{7}, -\frac{5\pi}{7}\}$, $\{-\frac{\pi}{7}, \frac{3\pi}{7}, \frac{5\pi}{7}\}$, then $\{\frac{\pi}{9}, \frac{5\pi}{9}, \frac{7\pi}{9}, -\frac{\pi}{9}, -\frac{5\pi}{9}, -\frac{7\pi}{9}\}$, etc.,...

A more elegant characterization, which we use in the theorem, uses the family $\{\delta_k\}_{k\geq 1}$ defined by

$$\delta_k = \gamma_k + \pi \qquad \text{modulo } 2\pi. \qquad (2.42)$$

Notice that $\gamma_{k+1} - \frac{\gamma_k \pm \pi}{2} \in 2\pi\mathbb{Z}$ implies that $\delta_k - 2\delta_{k+1} \in 2\pi\mathbb{Z}$. The set $\{\delta_n, \ldots, \delta_1\}$ in $[-\pi, \pi[$ is thus a cycle under the transformation $\omega \mapsto 2\omega$ modulo 2π. By Lemma 2.4, this is not the trivial cycle $\{0\}$. We can now state and prove the final result.

Theorem 2.2 *Let m_0 be a regular, 2π-periodic function that satisfies the hypotheses (2.1) and (2.2) for CQFs and that has only a finite number of zeros in the interval $[0, 2\pi]$. Then the following three properties are equivalent:*

i) The infinite product (2.3) generates a localized multiresolution analysis.

ii) There exists no non-trivial cycle $\{\delta_1, \ldots, \delta_n\}$ for the transformation $\omega \mapsto 2\omega$ modulo 2π, such that $|m_0(\delta_k)| = 1$.

iii) For every ω in $[-\pi, \pi]\setminus\{0\}$, the product $\prod_{k=0}^n m_0(2^k\omega)$ tends to 0.

Proof. Lemmas 2.1–2.5 have shown that ii)⇒ i). On the other hand, iii) ⇒ ii) follows directly since the existence of a non-trivial cycle implies that

$$\left|\prod_{k=1}^n m_0(2^k\delta_1)\right| = 1 \qquad \text{for all } n. \qquad (2.43)$$

It remains to show that i) ⇒ iii). For this, note that the hypothesis for a localized multiresolution analysis implies that the scaling function φ is in $L^1(\mathbb{R})$. Thus its Fourier transform $\hat{\varphi}(\omega)$ tends to 0 at infinity.

From the identity (2.18), for each ω in $[-\pi, \pi] \setminus \{0\}$, there exists an integer l such that $\hat{\varphi}(\omega+2l\pi)$ is not zero. Write $\omega_l = \omega+2l\pi \neq 0$. The expression

$$\hat{\varphi}(2^n\omega_l) = \prod_{k=0}^{n-1} m_0(2^k\omega_l)\,\hat{\varphi}(\omega_l) = \prod_{k=0}^{n-1} m_0(2^k\omega)\,\hat{\varphi}(\omega_l) \qquad (2.44)$$

must tend to 0 when n tends to $+\infty$. This is possible only if $\prod_{k=0}^{n-1} m_0(2^k\omega)$ tends to 0, which implies iii) and completes the proof of the theorem. $\square$

REMARKS

- In contrast with Theorem 2.1, which generalizes easily to several variables, this result appears to be difficult to extend to the multivariate setting. The zeros of the function m_0 are then algebraic curves and surfaces, which are more difficult to characterize. So far we do not know if i) and iii) are equivalent in several variables.

- It is important to remark that the hypothesis for the finite case is essential to establish Theorem 2.2. If it is abandoned, counterexamples appear immediately. It is sufficient to assume that m_0 vanishes on a small dyadic set of the type $[-2^{-j+1}, -2^{-j}] \cup [2^{-j}, 2^{-j+1}]$. If j is sufficiently large, this contradicts property (P) without there being a non-trivial cycle $\{\delta_1, \ldots, \delta_n\}$ on which $|m_0|$ is equal to 1.

We are now going to present a third characterization of the CQFs associated with orthonormal wavelets. This will bring into play the coefficients $\{h_n\}_{n \in \mathbb{Z}}$ that represent the impulse response of the filter m_0. We will study a family of conjugate quadrature filters and wavelets that are further restricted and more specific than those we have previously considered.

2.4 Wavelets with compact support

The theory of wavelets with compact support is presented in detail in Daubechies (1988, 1992). At the core of this theory is a construction based on the conjugate quadrature filters rather than on a given multiresolution analysis as, for example, in the case of spline wavelets (see Meyer (1990)). Wavelets with compact support have many interesting properties. They can be constructed to have a given number of derivatives and to have a given number of vanishing moments. What is more immediate, their associated filters have a finite impulse response and are particularly convenient for the implementation of the FWT. Indeed, if the functions φ and ψ have compact support, it is clear that the scalar products $\langle \varphi(\frac{x}{2}), \varphi(x-k) \rangle$ and $\langle \psi(\frac{x}{2}), \varphi(x-k) \rangle$, which define the coefficients of the associated CQF, are zero except for a finite number of the integers k.

The basic strategy for constructing φ and ψ is to design finite sequences $\{h_n\}_{n=N_0,\ldots,N_1}$ that satisfy (2.5) and (2.6) and to study the solution of the scaling equation (2.9). Indeed, assuming that the function m_0 is of the form

$$m_0(\omega) = \sum_{n=N_0}^{N_1} h_n e^{-in\omega}, \qquad (2.45)$$

we see that the function φ can be expressed, according to (2.3), by the convolution product

$$\varphi = \overset{+\infty}{\underset{k=1}{(*)}} \left(\sum_{n=N_0}^{N_1} h_n \delta(2^{-k} n) \right) = \overset{+\infty}{\underset{k=1}{(*)}} u_k, \qquad (2.46)$$

which converges in the sense of distributions. The support of the distribution u_k is $[2^{-k}N_0, 2^{-k}N_1]$, and, consequently, φ is a distribution with support in the interval $[N_0, N_1]$.

Noting that equation (2.1) is invariant when m_0 is multiplied by $e^{ik\omega}$, we see that it is possible to assume that our filter is of the form

$$m_0(\omega) = \sum_{n=0}^{N} h_n e^{-in\omega}. \qquad (2.47)$$

We then have a causal system, and the support of φ is in $[0, N]$.

The work of Ingrid Daubechies made it possible to classify all of the filters having finite impulse response in the case where the coefficients are real. In this case the scaling function and the wavelets are also real valued.

If N is even, then equation (2.5) implies that $h_0 \bar{h}_N = 0$, and this means that the filter must have an even number of coefficients. If we assume that the coefficients are real, then (2.5) implies that the filter is not symmetric, except in the case of the Haar system. In fact, the assumption that the filter is symmetric leads directly to the conclusion that $h_0 = h_N$ and $h_n = 0$ for all $n \neq 0$ or N.

The regularity of the distribution φ will be studied in the next chapter. For the time being, we mention that this distribution belongs to $L^2[0, N]$. This is a consequence of Fatou's lemma applied to the functions $\prod_{k=1}^{n} m_0(2^{-k}\omega) \mathbf{1}_{[-2^n\pi, 2^n\pi]}$, which, as we saw in the first section of this chapter, have a constant L^2 norm.

The setting we are in is more restrictive than the 'finite case,' and once again we will be concerned with the orthogonality of the sequence $\{\varphi(x - k)\}_{k \in \mathbb{Z}}$. The techniques that we are going to develop here owe a great deal to the ideas of Jean-Pierre Conze, Albert Raugi, and Wayne Lawton, which can be found in Conze and Raugi (1990) and Lawton (1991). We begin with a definition.

Definition 2.2 *Let $u(\omega)$ be a continuous function with period 2π. The transition operator P_u associated with u is defined by its action on the space of continuous, 2π-periodic functions:*

$$P_u f(\omega) = u\left(\frac{\omega}{2}\right) f\left(\frac{\omega}{2}\right) + u\left(\frac{\omega}{2} + \pi\right) f\left(\frac{\omega}{2} + \pi\right). \tag{2.48}$$

Notice that $\frac{\omega}{2}$ and $\frac{\omega}{2} + \pi$ are the two possible antecedents of ω under the transformation $\omega \mapsto 2\omega$ modulo 2π. Also, for future use, we write $S_0(\omega) = \frac{\omega}{2}$ and $S_1(\omega) = \frac{\omega}{2} + \pi$.

Now consider the finite dimensional vector space

$$E(N_1, N_2) = \left\{ \sum_{n=N_1}^{N_2} h_n e^{-in\omega} \Big| (h_{N_1}, \ldots, h_{N_2}) \in \mathbb{C}^{N_2 - N_1} \right\}. \tag{2.49}$$

If u and f belong to $E(N_1, N_2)$, then so does $P_u f$ because the function $u(\omega)f(\omega) + u(\omega + \pi)f(\omega + \pi)$ is in $E(2N_1, 2N_2)$ and its odd coefficients are zero. Thus if u is in $E(N_1, N_2)$, the study of P_u can be limited to this space. For the case that is of interest to us, we denote by P_0 the transition operator associated with the function $|m_0(\omega)|^2$. When m_0 is of the form (2.47), this function is in $E(-N, N)$, and we denote this space by E_N.

The matrix $M_0 = (M_{i,j})$ of P_0, as an operator on E_N, can be expressed conveniently as a function of the coefficients $h_0, \ldots, h_N$ of m_0. If we write $H_n = \overline{H}_{-n} = 2(h_n \overline{h}_0 + \cdots + h_N \overline{h}_{N-n})$, where the $\frac{1}{2} H_n$ are the Fourier coefficients of $|m_0(\omega)|^2$, we find that

$$M_{i,j} = H_{N+i-2j} \tag{2.50}$$

for $i, j = 0, \ldots, 2N$. Thus

$$M_0 = \begin{pmatrix} H_N & 0 & & \cdots & & 0 \\ H_{N-2} & H_{N-1} & H_N & 0 & \cdots & 0 \\ \vdots & & & & & \vdots \\ H_{-N} & H_{-N+1} & & \cdots & & H_N \\ 0 & 0 & H_{-N} & & \cdots & H_{N-2} \\ \vdots & & & & & \vdots \\ 0 & & & \cdots & 0 & H_{-N} \end{pmatrix}. \tag{2.51}$$

This particular structure is sometimes called a '2–slanted matrix.' It should be distinguished from the classic banded matrices that include the standard convolution operators and that cannot be studied in a finite dimensional space. In the present case, (2.48) shows that the convolution is followed by a decimation, which leads to this different structure.

From the relation (2.1), we note that P_0 satisfies

$$P_0(1) = 1. \tag{2.52}$$

Another invariant function of P_0 can be constructed from φ, although we do not know at this point whether φ generates a multiresolution analysis. It suffices to consider the series

$$\alpha(\omega) = \sum_{l \in \mathbb{Z}} |\hat{\varphi}(\omega + 2l\pi)|^2. \tag{2.53}$$

This is an element of E_N since its Fourier coefficients are given by

$$\alpha_n = \langle \varphi(x), \varphi(x-n) \rangle \tag{2.54}$$

and the support of φ is in $[0, N]$. Applying P_0 to α, we have

$$\begin{aligned}
P_0 \alpha(\omega) &= |m_0(\tfrac{\omega}{2})|^2 \sum_{l \in \mathbb{Z}} |\hat{\varphi}(\tfrac{\omega}{2} + 2l\pi)|^2 \\
&\quad + |m_0(\tfrac{\omega}{2} + \pi)|^2 \sum_{l \in \mathbb{Z}} |\hat{\varphi}(\tfrac{\omega}{2} + \pi + 2l\pi)|^2 \\
&= \sum_{l \in \mathbb{Z}} |\hat{\varphi}(\omega + 4l\pi)|^2 + \sum_{l \in \mathbb{Z}} |\hat{\varphi}(\omega + 2\pi + 4l\pi)|^2 \\
&= \alpha(\omega).
\end{aligned}$$

If the family $\{\varphi(x-k)\}_{k \in \mathbb{Z}}$ is orthonormal, then $\alpha(\omega) = 1$. This observation leads us to propose the following result.

Theorem 2.3 *The filter m_0 generates a multiresolution analysis and wavelets with compact support if and only if the only functions in E_N that are invariant under P_0 are the constants.*

Proof. The proof in one direction is almost evident from the remarks we have made. Indeed, suppose that the constants are the only invariant elements of P_0 in E_N. Then the function $\alpha(\omega)$ is a constant. But since $\alpha(0) = 1$, the result follows from Theorem 1.1. This shows that a sufficient condition to have a multiresolution analysis is that the characteristic polynomial $Q_0(x)$ of the matrix M_0 has a simple zero at 1, i.e. $Q_0(1) = 0$, and $Q_0'(1) \neq 0$.

The other direction of the proof is more difficult, and for this we will use the following definition.

WAVELETS WITH COMPACT SUPPORT

Definition 2.3 *We will say that a compact subset F of $[0, 2\pi]$ is invariant if and only if for all $\omega \in F$ and for all $\xi \in \{S_0, S_1\}$ such that $m_0(\xi(\omega)) \neq 0$, $\xi(\omega) \in F$.*

These sets were studied in detail by Conze and Raugi (1990). Here we only need a lemma that characterizes the finite invariant subsets.

Lemma 2.6 *A finite invariant subset F, associated with the function m_0, is the union of cyclic orbits of the transformation $\omega \mapsto 2\omega$, and $|m_0(\omega)| = 1$ for all $\omega \in F$.*

Proof. From (2.1), we can write

$$|m_0(S_0(\omega))|^2 + |m_0(S_1(\omega))|^2 = 1. \tag{2.55}$$

Consequently, if ω is an element of F, at least one member of the pair $\{S_0(\omega), S_1(\omega)\}$ is also an element of F. This allows us to construct a sequence of elements in F: We start with ω and apply successively the transformation S_0 or S_1.

These transformations are easily interpreted by considering the binary representation of $\frac{\omega}{2\pi}$, which is

$$\frac{\omega}{2\pi} = \sum_{j=1}^{+\infty} \omega_j\, 2^{-j} \qquad (\omega_j \in \{0, 1\}). \tag{2.56}$$

The action of S_0 is to add a 0 to the beginning of the sequence $\{\omega_j\}_{j \geq 1}$, and, similarly, S_1 adds a 1.

This implies that the binary expansion of $\frac{\omega}{2\pi}$ is periodic, for, otherwise, this process would generate an infinite number of points in F. (Recall a similar argument in Section 2.3.) Furthermore, the choice between S_0 or S_1 at each step in this process in uniquely determined by the periodic structure of the expansion. Thus, if $S_0(\omega)$ is in F, then $S_1(\omega)$ cannot belong to F, and we conclude that $m_0(S_1(\omega)) = 0$ and $|m_0(S_0(\omega))| = 1$. This is also true if S_0 and S_1 are interchanged, and it is true at each step in the process. Since the process must ultimately return us to the original point ω, we know that $|m_0(\omega)| = 1$. Note that the process we have described is exactly the inverse of iterating the transformation $\omega \mapsto 2\omega$ modulo 2π. This shows that F is the union of the cyclic orbits of its point under the transformation $\omega \mapsto 2\omega$ modulo 2π and that $|m_0(\omega)| = 1$ on these orbits. This completes the proof of the lemma. $\square$

Theorem 2.2 tells us that the existence of one of these cyclic orbits, when it is not $\{0\}$, implies that the construction of a multiresolution analysis is impossible. We finish the proof of Theorem 2.3 by constructing explicitly a finite invariant set in case the operator P_0 admits a non-constant invariant function in E_N.

Let $\beta(\omega)$ be such an invariant function (which we may assume is real), and let $B^m = \{b_1, \ldots, b_p\}$ be the set of points where $\beta(\omega)$ attains its global maximum in $[0, 2\pi[$. Then B^m is a finite invariant subset associated with m_0. To see this, write

$$\beta(b_j) = |m_0(S_0(b_j))|^2 \, \beta(S_0(b_j)) + |m_0(S_1(b_j))|^2 \, \beta(S_1(b_j)) \,. \quad (2.57)$$

Since

$$|m_0(S_0(\omega))|^2 + |m_0(S_1(\omega))|^2 = 1 \,,$$

$\beta(b_j)$ is a convex linear combination of $\beta(S_0(b_j))$ and $\beta(S_1(b_j))$. Consequently, $S_\varepsilon(b_j)$ is in B^m whenever $m_0(S_\varepsilon(b_j)) \neq 0$, and B^m is invariant.

By the same argument, the set of points B_m where $\beta(\omega)$ attains its global minimum on $[0, 2\pi[$ is also a finite invariant subset associated with m_0. At least one of the sets B^m and B_m does not reduce to $\{0\}$. Thus the assumption that P_0 has a non-constant invariant element contradicts the existence of a multiresolution analysis, and this completes the proof of Theorem 2.3. □

This theorem shifts the study of the filter m_0 to an investigation of the matrix M_0 given by (2.51). We note that the set of 'bad filters' is small in the sense that, if we fix the number of coefficients, these filters satisfy the extra condition $Q'_0(1) = 0$. Hence they form *a priori* a submanifold of codimension 1, which is a set of measure zero.

We conclude this chapter by presenting a final equivalence result that brings into play the FWT.

2.5 Action of the FWT on oscillating signals

In the preceding chapter, we said that m_0 was, in a weak sense, a 'low-pass' filter simply because its value at the origin was 1 and its value at π was 0. Theorem 2.2 allows us to justify this nomenclature more precisely. If it is associated with a multiresolution analysis, the filter m_0 eventually passes some high frequencies. However, its iterated application, combined with decimation at each step, tends

ACTION OF THE FWT ON OSCILLATING SIGNALS   59

to conserve only the zero frequency, as indicated by the property

$$\lim_{n\to+\infty} \prod_{k=0}^{n} m_0(2^k \omega) = 0 \quad \text{if} \quad \omega \in [-\pi, \pi] \setminus \{0\}. \tag{2.58}$$

This iterative mapping is used in the discrete wavelet decomposition to obtain the approximation at the scale 2^{-n}. Thus, our results can thus be interpreted in terms of the FWT.

To gain some intuition, we begin with some simple examples. Consider the periodic sequence $\{\ldots 0\,1\,0\,0\,1\,0\,0\,1\,0\,0\,1 \ldots\}$ and look at its image under the action of the Haar system. The first approximation is $\{\ldots \frac{1}{2}\,\frac{1}{2}\,0\,\frac{1}{2}\,\frac{1}{2}\,0\,\frac{1}{2}\,\frac{1}{2}\,0 \ldots\}$. At the second step we obtain $\{\ldots \frac{1}{4}\,\frac{1}{4}\,\frac{1}{2}\,\frac{1}{4}\,\frac{1}{4}\,\frac{1}{2}\,\frac{1}{4}\,\frac{1}{4}\,\frac{1}{2} \ldots\}$, and it is fairly clear that the process converges to the constant sequence $\{\frac{1}{3}\}$, which is nothing more than the average value of the original sequence.

On the other hand, if we use the filter $m_0(\omega) = \frac{1+e^{-3i\omega}}{2}$, which does not satisfy (2.58), the oscillations are conserved at each step because the original data is invariant under the action of m_0 and decimation.

Next, consider the sequence with zero mean value that is obtained by using the cube roots of unity $\{\ldots 1\,j\,j^2\,1\,j\,j^2\,1\,j\,j^2 \ldots\}$. This sequence is clearly still invariant under the action of the Haar filter dilated by a factor of three. It is also an eigenvector of the iterated Haar system since the second approximation gives $\{\ldots \frac{1}{4}\,\frac{j}{4}\,\frac{j^2}{4}\,\frac{1}{4}\,\frac{j}{4}\,\frac{j^2}{4} \ldots\}$. In this case, the property of being an 'eigenvector' is independent of the choice of m_0. It is the eigenvalue that varies with the choice of m_0. General sequences of this kind having period N can be constructed using the discrete Fourier transform. Thus, the discrete Fourier transform of the periodic sequence constructed from the Nth roots of unity is a 2π-periodic distribution consisting of Dirac masses at the points $\frac{2k\pi}{n}$.

We now look at a slightly different construction. Let $\{\omega_1, \ldots, \omega_n\}$ be a cycle for the mapping $\omega \mapsto 2\omega$ modulo 2π. We consider the sequence $\{s_k\}_{k\in\mathbb{Z}}$ defined by its discrete Fourier transform, which is $S(\omega) = \sum_{l\in\mathbb{Z}} \delta(\omega - \omega_1 + 2l\pi)$. Thus $s_k = e^{ik\omega_1}$, and since, by (2.56), $\omega_1 = 2\pi\frac{p}{q}$, it is clear that the sequence $\{s_k\}_{k\in\mathbb{Z}}$ is periodic.

Let m_0 be an arbitrary filter defined by $m_0(\omega) = \sum_{n\in\mathbb{Z}} h_n e^{-in\omega}$. The action of this filter on the sequence $\{s_k\}_{k\in\mathbb{Z}}$ is given by

$$(h * s)_k = \sum_{n\in\mathbb{Z}} h_{k-n} s_n = \sum_{n\in\mathbb{Z}} h_{k-n} e^{-in\omega_1} = m_0(\omega_1) s_k.$$

When this is followed by a decimation, which saves only the even terms, we get the sequence $\{s_{-1,k}\}_{k\in\mathbb{Z}}$ defined by

$$s_{-1,k} = m_0(\omega_1)s_{2k} = m_0(\omega_1)e^{i2k\omega_1} = m_0(\omega_1)e^{ik\omega_2}.$$

Passing to the discrete Fourier transform, we see that

$$s_{-1}(\omega) = m_0(\omega_1)\sum_{l\in\mathbb{Z}}\delta(\omega - \omega_2 + 2l\pi).$$

Thus, by the process of filtering and decimation, we pass from the pure frequency ω_1 to the pure frequency ω_2, plus multiplication by the scalar $m_0(\omega_1)$. When we repeat this operation n times to obtain the approximation at the resolution 2^{-n}, we get back the sequence $\{s_k\}_{k\in\mathbb{Z}}$ multiplied by the scalar $m_0(\omega_1)\cdots m_0(\omega_n)$. This means that we have found an eigenvector for the FWT approximation process iterated n times. These observations lead us naturally to the following result.

Theorem 2.4 *Let m_0 be a conjugate quadrature filter that satisfies the hypotheses of the 'finite case.'*

If it is associated with a multiresolution analysis, then the approximation of any discrete periodic signal tends toward its mean value as the resolution decreases.

If this is not the case, then there exists a non-constant periodic sequence that is equal, up to multiplication by a scalar of modulus 1, to its approximation at a lower resolution.

Proof. The proof follows almost directly from the preceding remarks. Let $\{s_k\}_{k\in\mathbb{Z}}$ be a periodic sequence with period n. Then its discrete Fourier transform is the periodic distribution given by

$$S(\omega) = \Big(\sum_{k=0}^{n-1} s_k e^{-ik\omega}\Big)\sum_{l\in\mathbb{Z}}\frac{1}{n}\delta\Big(\omega - \frac{2l\pi}{n}\Big).$$

If m_0 is a 'good low-pass filter,' it satisfies (2.58), and we have

$$\lim_{n\to+\infty}\prod_{k=0}^{n}m_0(2^k\omega)S(\omega) = \sum_{l\in\mathbb{Z}}S(0)\delta(2l\pi). \qquad (2.59)$$

This shows that the repeated action of $m_0(2^k\omega)$ destroys all the frequencies other than the fundamental. By taking the transform of both sides of (2.59), it is easy to show that the approximating sequences converge uniformly to the constant $\frac{1}{n}\sum_{k=o}^{n-1}s_k$. Indeed, it is a mater of the convergence of the product in (2.59) on a finite number of points.

If m_0 is a 'bad filter,' then by Theorem 2.2 we can find a cyclic orbit $\{\omega_1, \ldots, \omega_n\}$ for $\omega \mapsto 2\omega$ modulo 2π, which does not reduce to $\{0\}$, and such that $|m_0(\omega_i)| = 1$. Then choose the sequence $\{s_k\}_{k \in \mathbb{Z}}$ defined by $s_k = e^{ik\omega_1}$ and conclude, as before, that

$$\{s_{n,k}\}_{k \in \mathbb{Z}} = \left\{ \prod_{i=1}^{n} m_0(\omega_i)\, s_k \right\}_{k \in \mathbb{Z}}. \tag{2.60}$$

□

REMARKS

There is an analog for Theorem 2.4 for random signals. Conze and Raugi (1990) have shown that the application of the filter m_0, followed by a decimation, on a second-order stationary process can be expressed by the action of the transition operator P_0 on the power spectrum of the process. Recall that the power spectrum of a process s_n is defined by $p_s(\omega) = \sum_{k \in \mathbb{Z}} E(s_n s_{n+k}) e^{-ik\omega}$, where E stands for the expectation. Applying the results of Theorem 2.3, one can conclude that m_0 generates a multiresolution analysis if and only if the only second-order stationary signals that are invariant under the multiscale approximation are the constant processes.

This concludes our series of results on the equivalence between multiresolution analyses and the conjugate quadrature filters. We find these results interesting because of their diversity and because they give necessary and sufficient conditions. Nevertheless, their value for selecting filters for practical applications must be qualified. As we have indicated, the 'bad filters' constitute rare cases, and we will see in the following chapter why they are poorly adapted in most applications. The good filters remain to be classified. However, it must be noted that a large number of these so-called good filters hold little interest for practical applications. Our classification will involve the regularity of the functions φ and ψ associated with the CQF. These are properties that we have set aside until now. Our starting point will be the results of this chapter, which provide what seems to be the minimal requirements for the function m_0: the existence of a multiresolution analysis and the associated wavelet basis.

CHAPTER 3

The regularity of scaling functions and wavelets

3.1 Introduction

This chapter is devoted to studying the regularity of wavelets by using properties of their associated conjugate quadrature filters.

The ability to choose the function ψ so that it is well localized in both time and frequency (and hence very smooth) is one of the remarkable properties of wavelet bases. This is not the case, for example, for the families of the type $\{e^{inax}g(x-mb)\}_{m,n\in\mathbb{Z}}$. The Balian–Low theorem states that if such a family constitutes an orthonormal basis, then either the integral $\int(1+|x|)^2|g(x)|^2dx$ or $\int(1+|\omega|)^2|\hat{g}(\omega)|^2d\omega$ must diverge (see Daubechies (1992)). We further note that the classic cases of the Haar system and of the so-called 'Littlewood–Paley' wavelet, which is defined by its Fourier transform $\hat{\psi}(\omega) = \mathbf{1}_{[-2\pi,-\pi]} + \mathbf{1}_{[\pi,2\pi]}$, also fail to satisfy this double localization condition.

Added to the regularity and localization requirements is that of oscillation, or cancellation, which means that the function ψ has a certain number of zero moments. Endowed with these three properties, wavelets form unconditional bases for a large number of function spaces. Wavelets also allow one to characterize the global and local regularity of a function in terms of the decay properties of the coefficients in its wavelet expansion (see Meyer (1990)). In 1985, Yves Meyer constructed a wavelet that belongs to the Schwartz class and that has vanishing moments of all orders. Thus this function offers all of these advantages for functional analysis.

Our approach begins from a different point of view and poses additional questions:

- How can one characterize the regularity of the functions φ and ψ in terms of the properties of the CQFs?
- What can one expect in case the filters have compact support?
- What interest is there in the regularity of the wavelets from the

64 THE REGULARITY OF SCALING FUNCTIONS AND WAVELETS

strictly numerical point of view? Asked another way, what is the implication of regularity for the FWT algorithm?

We answer these questions using a spectral approach. This means that we use a frequency localization criterion to characterize the regularity of the functions φ and ψ. Here we extend the investigations of Daubechies (1988).

Our spectral criterion, by its nature, does not provide the kind of exact, local information about the regularity of φ and ψ that can be obtained with a spatial approach (see Daubechies and Lagarias (1991, 1992)). Nevertheless, it has the advantage of giving us, quite easily, bounds on the global Hölder exponent of these functions and an asymptotic result about the regularity of wavelets associated with long filters.

Throughout this chapter, we assume the hypotheses for localized multiresolution analyses as announced in Chapter 1. In particular, recall that we have at hand the functions φ, ψ, m_0, and m_1, and that they satisfy these key relations:

$$\int (1+|x|)^m |\varphi(x)|^2 dx < +\infty \quad \text{for all} \quad m \in \mathbb{N}, \qquad (3.1)$$

$$m_0(\omega) = \sum_{k \in \mathbb{Z}} h_k e^{-ik\omega}, \quad \text{where the } h_k \text{ decrease rapidly},$$

$$m_1(\omega) = e^{-i\omega} \overline{m_0(\omega + \pi)} = \sum_{k \in \mathbb{Z}} g_k e^{-ik\omega} = \sum_{k \in \mathbb{Z}} (-1)^k \overline{h_{1-k}} e^{-ik\omega},$$

$$\frac{1}{2} \varphi(\frac{x}{2}) = \sum_{k \in \mathbb{Z}} h_k \, \varphi(x-k),$$

and

$$\frac{1}{2} \psi(\frac{x}{2}) = \sum_{k \in \mathbb{Z}} g_k \, \varphi(x-k). \qquad (3.2)$$

3.2 Regularity and oscillation

It is clear from (3.2) that the regularity properties of the wavelet ψ are inherited directly form those of the scaling function φ because the coefficients $\{g_n\}_{n \in \mathbb{Z}}$ decrease rapidly.

A first 'regularity scale' or 'smoothness scale' can be introduced in terms of the following definition.

Definition 3.1 *Let r be a positive integer. We say that a multiresolution analysis is 'r-regular' if and only if the scaling function satisfies the following three properties:*

- φ is $r-1$ times continuously differentiable.
- φ is r times differentiable almost everywhere.
- For all m in $\mathbb{N}$ and for all n such that $0 \leq n \leq r$, we have

$$\int (1+|x|)^m \left|\left(\frac{d}{dx}\right)^n \varphi(x)\right|^2 dx < +\infty. \tag{3.3}$$

Another way to say this is that the derivatives of φ up to order r satisfy the localization property.

The prototype for the r-regular multiresolution analyses is given by

$$V_j = \left\{ f \in L^2(\mathbb{R}) \cap \mathcal{C}^{r-1}(\mathbb{R}) \ \Big|\ \left(\frac{d}{dx}\right)^{r+1} f = \sum_{k \in \mathbb{Z}} \alpha_k\, \delta(2^{-j}k) \right\}. \tag{3.4}$$

These are the spline functions of degree r on the dyadic intervals $]2^{-j}k, 2^{-j}(k+1)[$. One obtains the spline orthonormal wavelets from this system. These wavelets do not have compact support, but they do decrease exponentially (see Meyer (1990)).

The following result, which is due to Yves Meyer, shows that the oscillation property arises naturally from the regularity.

Proposition 3.1 *Let ψ be a wavelet that comes from an r-regular multiresolution analysis. Then, for $0 \leq n \leq r$,*

$$\left(\frac{d}{d\omega}\right)^n \hat{\psi}(0) = \int x^n \psi(x)\, dx = 0. \tag{3.5}$$

Proof. Note that there always exists a point x_0 where the function ψ is r times differentiable and such that the numbers $\psi(x_0)$, $\psi'(x_0), \ldots, \psi^r(x_0)$ are all different from zero. If this were not the case, we could deduce from the regularity hypothesis that ψ is identically zero. Note also that all of the moments of ψ exist because of the localization hypothesis (3.1).

For each j in $\mathbb{Z}$, we define k_j to be the integer that satisfies the inequality $2^{-j} k_j \leq x_0 < 2^{-j}(k_j+1)$. We are going to use the wavelets ψ_{j,k_j} that are localized around x_0 when j is large. Thus, for $j > 0$, we write

$$0 = \langle \psi, \psi_{j,k_j} \rangle = 2^{-j/2} \langle \psi_{j,k_j}, \psi \rangle = \int \overline{\psi(x)}\, \psi(2^j x - k_j)\, dx$$

$$= \int \Big[\overline{\psi(x_0)} + (x-x_0)\overline{\psi'(x_0)} + \cdots$$

$$\cdots + (x-x_0)^r \frac{\overline{\psi^r(x_0)}}{r!}(1+\alpha(x-x_0))\Big] \psi(2^j x - k_j)\, dx.$$

Here we have used a limited development of $\psi(x)$ about the point x_0. The function α is bounded, continuous at 0, and such that $\alpha(0) = 0$.

By making the change of variable $x \mapsto 2^{-j}x + x_0$, we see that

$$0 = 2^{-j}\overline{\psi(x_0)} \int \psi(x + 2^j x_0 - k_j)\,dx$$
$$+ 2^{-2j}\overline{\psi'(x_0)} \int x\,\psi(x + 2^j x_0 - k_j)\,dx + \cdots$$
$$+ 2^{-(r+1)j}\frac{\overline{\psi^r(x_0)}}{r!} \int x^r(1 + \alpha(2^{-j}x))\,\psi(x + 2^j x_0 - k_j)\,dx.$$

Note that $|2^j x_0 - k_j| \leq 1$. This, combined with the fact that ψ inherits the condition (3.1) from φ, implies that all of the integrals $\int x^n \psi(x + 2^j x_0 - k_j)\,dx$ (including the last one) are uniformly bounded for $0 \leq n \leq r$ and $j \in \mathbb{Z}$. Now replace x by $x - 2^j x_0 + k_j$ in the first integral, multiply both sides by 2^j, and let j tend to $+\infty$. All the terms except the first one tend to zero, and we are left with $0 = \overline{\psi(x_0)} \int \psi(x)\,dx$, which, by the choice of x_0, means that $\int \psi(x)\,dx = 0$.

Using the same process—changing variable in the first remaining integral and multiplying by 2^j—we conclude that $\int x^n \psi(x)\,dx = 0$ for $0 \leq n \leq r - 1$. The induction step is illustrated by dealing with the last integral. Thus assume that the first r moments are zero. Then the above relation collapses to

$$0 = \int x^r(1 + \alpha(2^{-j}x))\,\psi(x + 2^j x_0 - k_j)\,dx.$$

When we replace x by $x - 2^j x_0 + k_j$, we get

$$0 = \int (P_{r-1}(x) + x^r)(1 + \alpha(2^{-j}x - x_0 + 2^{-j}k_j))\,\psi(x)\,dx,$$

where P_{r-1} is a polynomial of degree at most $r-1$. By assumption, $\int P_{r-1}(x)\,\psi(x)\,dx = 0$ for each j, and we are left with

$$0 = \int x^r\,\psi(x)\,dx + \int (P_{r-1}(x) + x^r)\alpha(2^{-j}x - x_0 + 2^{-j}k_j)\,\psi(x)\,dx.$$

The function in the second integral is bounded by

$$(1 + |x|)^r |\alpha(2^{-j}x - x_0 + 2^{-j}k_j)|\,|\psi(x)|.$$

Recalling that $|x_0 - 2^{-j}k_j| \leq 2^{-j}$, we see that this function tends to zero for each fixed x when j tends to $+\infty$. This function is, in turn, bounded by an integrable function, so the second integral tends to zero by dominated convergence. We are left with $\int x^r \, \psi(x)\, dx = 0$, and this proves the proposition. □

This result shows that the regularity up to order r forces the first $r+1$ moments to vanish. One can ask, conversely, if this last property implies that the wavelets are in $\mathcal{C}^r$ or $\mathcal{C}^{r-1}$. We will see shortly that this is not the case.

We note that regularity and oscillation are complementary properties. Cancellation is essential for analysis because it is cancellation that causes the scalar products of wavelets with smooth functions to be very small for large j, that is, at fine scales. On the other hand, regularity is useful in the synthesis because it allows the precise reconstruction of these same functions, in the sense that the wavelet expansion converges in the smoothness space where the function belongs. This is the case, for example, when the function belongs to one of the Hölder or Sobolev spaces.

Finally, we note that the property of having vanishing moments is easily interpreted from the numerical point of view. Indeed, in a neighborhood of the origin

$$m_1(\omega) = e^{-i\omega}\overline{m_0(\omega+\pi)} = \frac{\hat{\psi}(2\omega)}{\hat{\varphi}(\omega)}, \qquad (3.6)$$

which means that (3.5) is equivalent to

$$\left(\frac{d}{d\omega}\right)^n m_1(0) = \sum_{k\in\mathbb{Z}} k^n g_k = 0 \quad \text{when} \quad 0 \le n \le r. \qquad (3.7)$$

The high-pass filter m_1, which is used to calculate the wavelet coefficients at each scale from the approximations, satisfies the same oscillation properties as ψ. The vanishing of the first $r+1$ moments means that the analysing object is sensitive to the size of the derivative of order $r+1$. The action of the filter can thus be interpreted as a discrete derivation of order $r+1$.

In contrast, the effect of the regularity properties of the filters on the algorithms seems difficult to describe, in particular, because one is dealing with discretized functions for which the concept of smoothness has no meaning. To understand better the numerical significance of regularity, we will return to the process by which we constructed a scaling function from a conjugate quadrature filter.

3.3 The subdivision algorithms

Up to now we have used the product

$$\hat{\varphi}(\omega) = \prod_{k=1}^{+\infty} m_0(2^{-k}\omega) \qquad (3.8)$$

to relate the scaling function to the CQF. Reexamining the formula for changing scale in the spatial domain, that is, the relation

$$\varphi(x) = 2 \sum_{n \in \mathbb{Z}} h_n \varphi(2x - n), \qquad (3.9)$$

we observe that the function φ is a fixed point for the operator

$$Tf(x) = 2 \sum_{n \in \mathbb{Z}} h_n f(2x - n). \qquad (3.10)$$

Thus we can try to construct the function φ by iterating this operator on a suitable initial function.

We note that the property

$$\sum_{n \in \mathbb{Z}} h_n = 1 \qquad (3.11)$$

implies the invariance under T of the integral (if it exists), which is to say that

$$\int Tf(x)\, dx = \int f(x)\, dx. \qquad (3.12)$$

We take for our initial element one of the simplest functions whose integral is 1, namely, the piecewise affine 'hat function' u_0 defined by $u_0(x) = \max\{0, 1 - |x|\}$

At step p, we obtain a function $u_p = T^p u_0$ that is piecewise affine on each interval $[2^{-p}k, 2^{-p}(k+1)]$, as illustrated in Figures 3.1–3.4, which will be described later. The Fourier transform of this function is given by

$$\hat{u}_p(\omega) = \prod_{k=1}^{p} m_0(2^{-k}\omega)\, \hat{u}_0(2^{-p}\omega). \qquad (3.13)$$

In a moment, we will examine the convergence of the sequence u_p to φ, but first we note that these functions, $u_p = T^p u_0$, are completely determined by their samples $s_{p,k} = u_p(2^{-p}k)$. This leads us to examine how the operator T transforms the sequences $\{s_{p,k}\}_{k \in \mathbb{Z}}$ as p increases. By using (3.10), the recursion relation becomes

$$s_{p+1,k} = u_{p+1}(2^{-p-1}k) = 2\sum_{n\in\mathbb{Z}} h_n\, u_p(2^{-p}k - n)$$
$$= 2\sum_{n\in\mathbb{Z}} h_n\, s_{p,k-2^p n}\,. \tag{3.14}$$

Define S_p by

$$S_p(\omega) = \sum_{k\in\mathbb{Z}} s_{p,k}\, e^{i2^{-p}k\omega}\,, \tag{3.15}$$

which is the discrete Fourier transform of $\{s_{p,k}\}_{k\in\mathbb{Z}}$ considered as an element of $l^2(2^{-p}\mathbb{Z})$. In the frequency domain, (3.14) becomes

$$S_{p+1}(\omega) = 2m_0\left(\frac{\omega}{2}\right) S_p\left(\frac{\omega}{2}\right). \tag{3.16}$$

Since $S_0(\omega) = 1$, this last equation shows that

$$S_p(\omega) = 2^p \prod_{k=1}^{p} m_0(2^{-k}\omega)\,. \tag{3.17}$$

Looked at slightly differently, this says that

$$S_{p+1}(\omega) = 2S_p(\omega)\, m_0(2^{-p-1}\omega)\,, \tag{3.18}$$

which is interpreted in the spatial domain as

$$s_{p+1,n} = 2\sum_{k\in\mathbb{Z}} h_{n-2k}\, s_{p,k}\,. \tag{3.19}$$

This new formula is more interesting than (3.14), and it is the one that we used in Figures 3.1–3.4. It shows that our algorithm has a local character, which is not evident from (3.14). Specifically, equation (3.19) shows that the sequence $\{s_{p,k}\}_{k\in\mathbb{Z}}$ is refined into $\{s_{p+1,k}\}_{k\in\mathbb{Z}}$ in such a way that the value of u_p at the point $2^{-p}k$ influences the value of u_{p+1} only on the neighborhood $[2^{-p}k + 2^{-p-1}N_1, 2^{-p}k + 2^{-p-1}N_2]$ if $h_n = 0$ when n is outside the neighborhood $[N_1, N_2]$.

With this different point of view, this 'cascade algorithm,' which was introduced by Ingrid Daubechies in (1988), can be identified with methods of 'subdivision' that have been widely studied in the theory of approximation of curves and surfaces and in computer aided geometric design.

Generally speaking, subdivision algorithms are designed for the fast generation of smooth curves and surfaces from a sparse set of control points by means of iterative refinements. The 'binary univariate' case is the one that is most often considered. Here, one starts with a sequence $\{s_0(k)\}_{k\in\mathbb{Z}}$ and obtains at step j a sequence $\{s_j(2^{-j}k)\}_{k\in\mathbb{Z}}$ that is generated from the previous one by a linear rule of the form

$$s_j(2^{-j}k) = 2 \sum_{n\in k+2\mathbb{Z}} h_{j,k,n}\, s_{j-1}(2^{-j+1}(k-n))\,. \qquad (3.20)$$

The 'masks' $h_{j,k} = \{h_{j,k,n}\}_{k\in\mathbb{Z}}$ are generally finite sequences, a property that is clearly useful for the practical implementation of (3.20).

A natural problem is then to study the convergence of such an algorithm to a limit function. For this, we say that the subdivision scheme is strongly (or uniformly) convergent if and only if there exists a continuous function f such that

$$\lim_{j\to+\infty} \left[\sup_{k\in\mathbb{Z}} |s_j(2^{-j}k) - f(2^{-j}k)| \right] = 0\,.$$

If f is differentiable, we can define

$$s'_j(2^{-j}k) = 2^j [s_j(2^{-j}k) - s_j(2^{-j}(k-1))]$$

and ask if

$$\lim_{j\to+\infty} \left[\sup_{k\in\mathbb{Z}} |s'_j(2^{-j}k) - f'(2^{-j}k)| \right] = 0\,.$$

This can of course be extended to higher derivatives.

A subdivision scheme is said to be stationary and uniform when the masks $h_{j,k}$ are independent of j and k. In this case we can rewrite (3.20) as

$$s_j(2^{-j}k) = 2 \sum_{n\in\mathbb{Z}} h_{k-2n}\, s_{j-1}(2^{-j+1}n) \qquad (3.21)$$

which is exactly the form of (3.19). A detailed review of subdivision algorithms and their possible generalizations can be found in Deslauriers and Dubuc (1987), Dyn (1992), and Cavaretta, Dahmen, and Micchelli (1991).

Finally, we remark that (3.18) is exactly the reconstruction formula in the FWT algorithm if one keeps only the approximation and not the details (see (1.39) in Chapter 1). Iterating the subdivision p times thus produces the elementary component of the discrete signal corresponding to the coefficients for approximation at the resolution 2^p. The component corresponding to the coefficients for the details is obtained by applying the high-pass filter m_1 once, either at the first step of the algorithm if one uses (3.19) or at the last step if one uses (3.14). One can identify this with the discrete wavelet introduced in Chapter 1.

In many applications, such as the compression and coding discussed in Chapter 5, one hopes that these components have an aspect of continuity or differentiability. Since they approximate the functions φ and ψ, it will be interesting, from the strictly numerical point of view, to study the regularity of these functions.

We next look at four examples of applications of the subdivision algorithm to get a better idea of the problems related to regularity. Each of the Figures 3.1, 3.2, 3.3, and 3.4 represents the approximation of φ at steps $p = 1, 4$, and 7.

In Figure 3.1, the CQF used corresponds to a wavelet of the class $\mathcal{C}^2$. Notice that the regularity is apparent at step $p = 4$.

Figure 3.2 illustrates a case where the convergence is poor. Although the CQF satisfies the hypotheses of the theorems of the preceding chapter and generates a multiresolution analysis, the wavelets enjoy very little regularity and are thus of little interest, either from the mathematical point of view or from the numerical point of view.

Figure 3.3 shows the result of a subdivision that uses the Haar system dilated by a factor of 3. Recall that this CQF does not generate a multiresolution analysis. We see that this dilation introduces 'gaps,' thus creating an irregularity that becomes more and more pronounced. We note however that this subdivision converges to the function $\frac{1}{3}u_0\left(\frac{x}{3}\right)$ (which satisfies (3.9)) but only in the sense of distributions.

This bad behavior can be understood from Theorem 2.2. Specifically, the discrete Fourier transform $S_p(\omega)$ contains high frequencies that do not become blurred when p tends to $+\infty$, which explains the oscillatory behavior.

Finally, we see in Figure 3.4, which also corresponds to a counterexample, that the limit function of the subdivision taken in the sense of distributions can be regular while the convergence remains poor. This is surely not a desirable situation because, from the numerical point of view, the successive iterations are more important than the final result. Thus we look for a criterion that ensures the regularity of φ and, at the same time, guarantees 'good' convergence for the algorithm. The following result provides a first step in this direction.

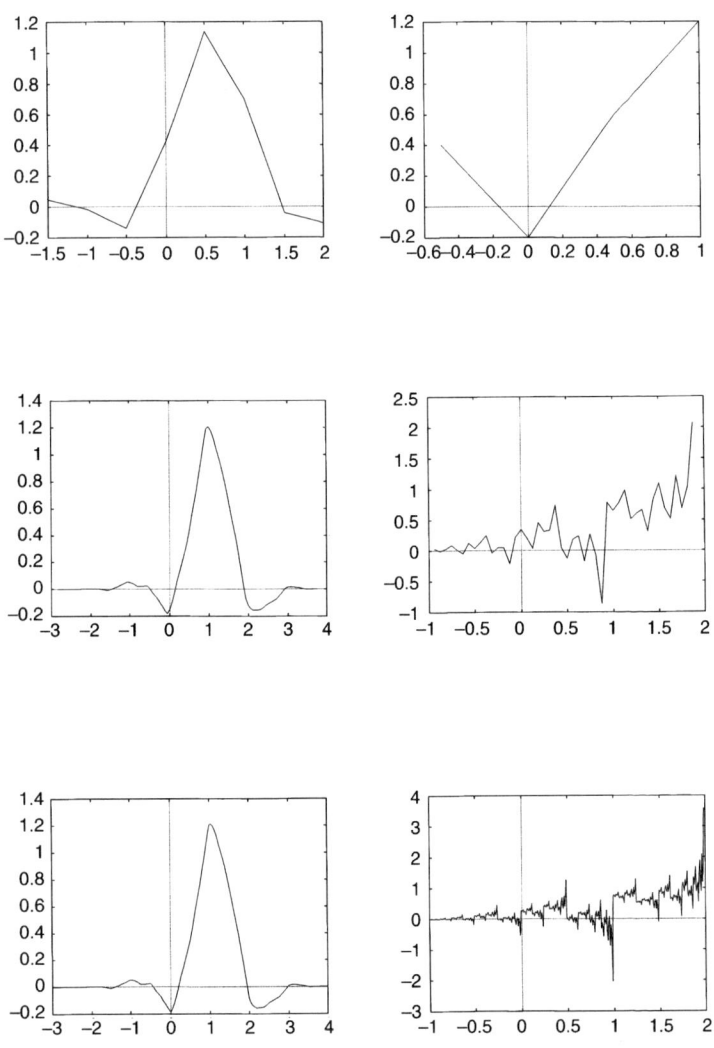

Figure 3.1 Figure 3.2

Examples of subdivisions producing orthonormal bases. The convergence is at least in $L^2(\mathbb{R})$.

THE SUBDIVISION ALGORITHMS

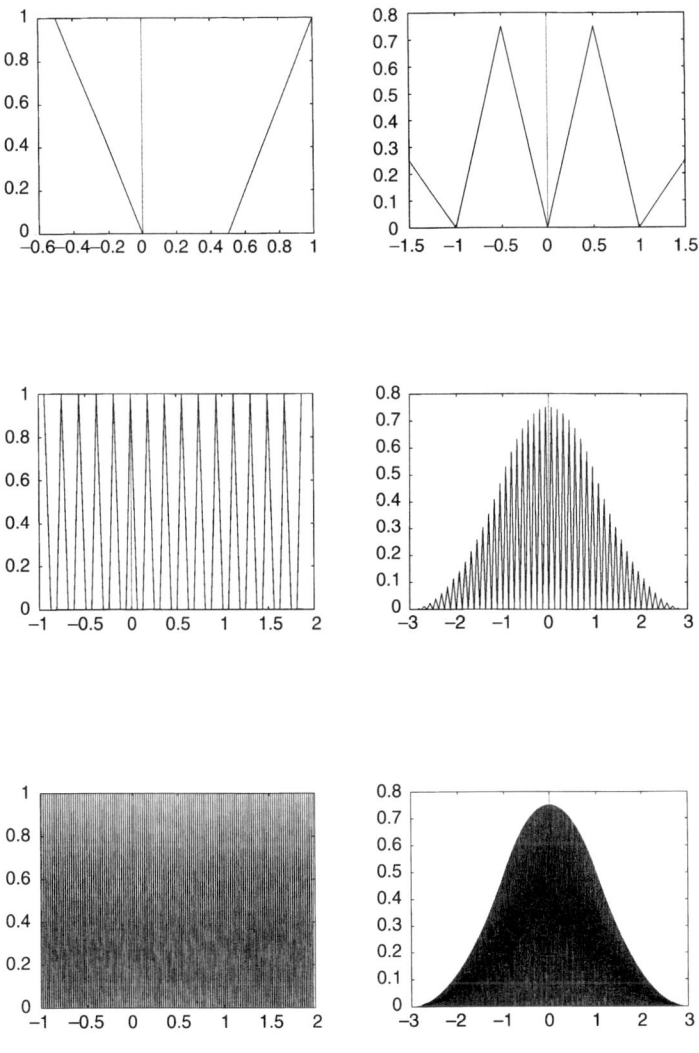

Figure 3.3 Figure 3.4

*Subdivisions associated with CQFs not satisfying (P).
The convergence is only in the sense of distributions.*

74 THE REGULARITY OF SCALING FUNCTIONS AND WAVELETS

Proposition 3.2 *Suppose that the filter m_0 generates a multiresolution analysis, which is to say that it satisfies the hypotheses of Theorem 2.1. Furthermore, assume that*

$$\int (1 + |\omega|^r) |\hat{\varphi}(\omega)| \, d\omega < +\infty \tag{3.22}$$

for some positive integer r. Then the function φ is r times differentiable, and the subdivision algorithm converges uniformly to φ, as do all of the discrete derivatives up to order r.

Proof. The regularity of φ is an immediate consequence of (3.22). To prove the convergence, we go back to the proof of Theorem 2.1. There we used the sequence of functions $\{f_n\}_{n \geq 0}$ defined by

$$\hat{f}_n(\omega) = \prod_{k=1}^{n} m_0(2^{-k}\omega) \, \mathbf{1}_{2^n K}, \tag{3.23}$$

where K is a compact set congruent to $[-\pi, \pi]$ modulo 2π that satisfies property (P). Reasoning analogous to that of Theorem 2.1 leads to the inequality

$$|\omega|^q |\hat{f}_n(\omega)| \leq \frac{|\omega|^q |\hat{\varphi}(\omega)|}{C}. \tag{3.24}$$

We can now apply Lebesgue's dominated convergence theorem and conclude that the sequences $\omega^q \hat{f}_n(\omega)$ converge to the limits $\omega^q \hat{\varphi}(\omega)$ in L^1 for $q \leq r$. Consequently, the functions f_n converge uniformly to φ, and the derivatives of the f_n, up to order r, converge uniformly to those of φ.

Next, we need to relate the functions f_n to the functions u_n and to the sequences in the subdivision algorithm. The result is simply that

$$s_{n,k} = u_n(2^{-n}k) = f_n(2^{-n}k). \tag{3.25}$$

This can be seen by computing the Fourier transform of $\hat{f}_n$ using the properties of K exactly as we did in the proof of Theorem 2.1. This computation shows that

$$f_n(2^{-n}k) = 2^n \frac{1}{2\pi} \int_{-\pi}^{\pi} \prod_{l=0}^{n-1} m_0(2^l \omega) e^{ik\omega} \, d\omega. \tag{3.26}$$

Equation (3.25) follows directly from (3.26) and

$$S_n(2^n \omega) = 2^n \prod_{l=0}^{n-1} m_0(2^l \omega) = \sum_{k \in \mathbb{Z}} s_{n,k} \, e^{ik\omega} \tag{3.27}$$

THE SUBDIVISION ALGORITHMS 75

because the $f_n(2^{-n}k)$ are the Fourier coefficients of $S_n(2^n\omega)$, which are in turn the $s_{n,k}$.

Equation (3.25) shows that the result of the algorithm at step n corresponds exactly to sampling the function f_n uniformly at the points $2^{-n}k$. This means that the algorithm converges uniformly to φ, which is to say that the subdivision scheme $s_{n,k}$ converges uniformly to φ as n tends to $+\infty$.

We now examine the convergence of the discrete derivatives of order $q \leq r$. These derivatives are defined by

$$s_{n,k}^{(q)} = \frac{1}{2^{-nq}} \sum_{p=0}^{q} b_p \, s_{n,k+p}, \qquad (3.28)$$

where b_p is the coefficient of x^p in $(x-1)^q$.

To prove the uniform convergence of the $s_{n,k}^{(q)}$ to $\left(\frac{d}{dx}\right)^q \varphi$, we show that

$$|s_{n,k}^{(q)} - \left(\frac{d}{dx}\right)^q f_n(2^{-n}k)| \qquad (3.29)$$

is small uniformly in k for all sufficiently large n.

We apply the difference operator in (3.28) to the f_n, with $2^{-n}k$ replaced by x and 2^{-n} replaced by h, and write

$$\Delta^{(q)} f_n(x,h) = \frac{1}{h^q} \sum_{p=0}^{q} b_p f_n(x+ph). \qquad (3.30)$$

These differences should approximate the derivatives $\left(\frac{d}{dx}\right)^q f_n$. In fact,

$$\Delta^{(q)} f_n(x,h) - \left(\frac{d}{dx}\right)^q f_n(x)$$
$$= \frac{1}{2\pi} \int (i\omega)^q \hat{f}_n(\omega) e^{ix\omega} \left[\left(\frac{e^{ih\omega}-1}{ih\omega}\right)^q - 1\right] d\omega. \qquad (3.31)$$

By (3.24), the integral on the right-hand side is dominated by

$$C \int (1+|\omega|^r)|\hat{\varphi}(\omega)| \left|\left(\frac{e^{ih\omega}-1}{ih\omega}\right)^q - 1\right| d\omega \qquad (3.32)$$

uniformly in x and n, where the constant C can be chosen to be independent of these parameters and of q ($q \leq r$). Thus, by dominated convergence, $\Delta^{(q)} f_n(x,h)$ tends to $\left(\frac{d}{dx}\right)^q f_n(x)$ uniformly in x and the other parameters when $|h|$ tends to zero. In particular, since $\Delta^{(q)} f_n(2^{-n}k, 2^{-n}) = s_{n,k}^{(q)}$, $s_{n,k}^{(q)}$ tends to $\left(\frac{d}{dx}\right)^q f_n(2^{-n}k)$ uniformly in k as n tends to infinity. To recapitulate, we have shown that the $s_{n,k}^{(q)}$ become uniformly close to the $\left(\frac{d}{dx}\right)^q f_n(2^{-n}k)$ as n tends to $+\infty$, while we know that the $\left(\frac{d}{dx}\right)^q f_n$ tend uniformly to $\left(\frac{d}{dx}\right)^q \varphi$. This completes the proof of Proposition 3.2. □

REMARK

It is clear from the examples in Figures 3.4 and 3.5 that property (P) of Theorem 2.1 is essential to achieve the result we have just proved. The orthonormality conditions thus appear to be the minimum that are required for numerical applications, and we will assume these conditions in the remainder of the chapter. We must now characterize those CQFs for which the associated scaling function φ satisfies the frequency localization property (3.22).

3.4 Spectral estimates of the regularity

The regularity of the function φ can be measured in different ways. We will discuss, in particular, the global Hölder exponent: If φ is in $\mathcal{C}^n$ but not in $\mathcal{C}^{n+1}$, then its Hölder exponent is given by $\mu = n+\nu$, where ν is the supremum of the $\alpha \in [0,1[$ such that $\varphi^{(n)} \in \mathcal{C}^\alpha$, or equivalently,

$$\nu = \inf_x \left(\liminf_{|t| \to 0} \frac{\log |\varphi^{(n)}(x+t) - \varphi^{(n)}(x)|}{\log |t|} \right).$$

In the context of the scaling functions that are defined by the infinite product (3.8), we will also be interested in a smoothness measurement based on the Fourier transform. For $1 \leq p \leq +\infty$, we introduce the L^p-Sobolev exponent

$$s_p(\varphi) = \sup\{\alpha \mid (1+|\omega|^\alpha)\hat\varphi(\omega) \in L^p\}.$$

The case $p=2$ coincides with the standard Sobolev exponent. It is also clear that $\mu \geq s_1$. Furthermore, by Hölder's inequality, one can derive the following relation:

$$s_q \geq s_p + \frac{1}{p} - \frac{1}{q} \qquad \text{for} \qquad q \leq p.$$

In particular, this relation holds for $p = +\infty$ so that $s_q + \frac{1}{q} \geq s_\infty$ and $\mu \geq s_1 \geq s_\infty - 1$.

3.4.1 A condition sufficient for regularity

A condition that implies the decay of the Fourier transform of φ was introduced by Ingrid Daubechies (1988). It is based essentially on the nature of the zero of the function m_0 at π. Suppose, in addition to having $m_0(\pi) = 0$, we know that m_0 can be factored in the form

$$m_0(\omega) = \left(\frac{1+e^{i\omega}}{2}\right)^N p(\omega), \tag{3.33}$$

where p is in $\mathcal{C}^\infty(\mathbb{R})$ and is 2π-periodic. This is true, in particular, when m_0 is a trigonometric polynomial whose first $N-1$ derivatives vanish at π. Note from (3.6) that N is also the degree of the zero of $\hat{\psi}$ at the origin, which we have seen (Proposition 3.1) is related to the regularity of ψ.

For $j \geq 0$, define

$$B_j = \sup_{\omega \in \mathbb{R}} \left| \prod_{k=1}^{j} p(2^{-k}\omega) \right|, \tag{3.34}$$

and let $B_0 = 1$. For $j > 0$, we write

$$b_j = \frac{1}{j \log 2} \log B_j. \tag{3.35}$$

A few observations are useful for what follows. Because the supremum is taken over $\mathbb{R}$,

$$B_j = \sup_{\omega \in \mathbb{R}} \left| \prod_{k=1}^{j} p(2^{-k-m}\omega) \right|$$

for any m in $\mathbb{Z}$. Also, since $m_0(0) = 1$, $p(0) = 1$, and hence $B_j \geq 1$ and $b_j \geq 0$.

If j and n are positive integers, we can write $n = mj + r$ where $m \geq 0$ and $0 \leq r < j$. With this notation, we have the following inequalities:

$$B_n \leq (B_j)^m B_r \leq (B_j)^m (B_1)^r. \tag{3.36}$$

Finally, note that, with the assumptions we have made, the product $\prod_{k=1}^{+\infty} p(2^{-k}\omega)$ converges on compact sets to a smooth function. Thus, for each $A > 0$ there is a constant Q_A such that

$$\sup_{n \geq 1} \sup_{|\omega| \leq A} \prod_{k=1}^{n} |p(2^{-k}\omega)| = Q_A < +\infty. \tag{3.37}$$

78 THE REGULARITY OF SCALING FUNCTIONS AND WAVELETS

Proposition 3.3 *With the preceding definitions, for each $j > 0$, there exists a constant C_j such that*

$$|\hat{\varphi}(\omega)| = \left|\prod_{k=1}^{+\infty} m_0(2^{-k}\omega)\right| \leq C_j(1+|\omega|)^{-N+b_j} \quad (3.38)$$

for all ω in $\mathbb{R}$.

Proof. Note that $\left|\frac{1+e^{i\omega}}{2}\right| = \left|\cos\left(\frac{\omega}{2}\right)\right|$. We use the classical identity

$$\prod_{k=1}^{+\infty} \cos(2^{-k}\omega) = \frac{\sin \omega}{\omega} \quad (3.39)$$

and the inequality $\left|\frac{\sin \omega/2}{\omega/2}\right| \leq 4(1+|\omega|)^{-1}$ to obtain the estimate

$$|\hat{\varphi}(\omega)| \leq 4^N(1+|\omega|)^{-N} \prod_{k=1}^{+\infty} |p(2^{-k}\omega)|. \quad (3.40)$$

To estimate the second infinite product, fix ω and let n be the unique positive integer such that $2^n \leq 1+|\omega| < 2^{n+1}$. Since $|2^{-n}\omega| < 2$ we have

$$\prod_{k=1}^{+\infty} |p(2^{-k}\omega)| \leq \prod_{k=1}^{n} |p(2^{-k}\omega)| \prod_{k=1}^{+\infty} |p(2^{-k-n}\omega)|$$
$$\leq B_n Q_2.$$

To estimate B_n, write $n = mj + r$ where $m \geq 0$ and $0 \leq r < j$. Extending (3.36), we have

$$B_n \leq (B_j)^m (B_1)^r \leq (B_j)^{n/j}(B_1)^j.$$

The inequality $\frac{n}{j} \leq \frac{\log(1+|\omega|)}{j \log 2}$ and the definition of b_j are used to estimate $(B_j)^{n/j}$:

$$(B_j)^{n/j} \leq (B_j)^{\frac{\log(1+|\omega|)}{j \log 2}} = (1+|\omega|)^{b_j}. \quad (3.41)$$

Putting all of this together, with $C_j = 4^N Q_2 (B_1)^j$, we conclude that

$$|\hat{\varphi}(\omega)| \leq C_j (1+|\omega|)^{-N+b_j} \quad (3.42)$$

for all ω in $\mathbb{R}$, and this proves the proposition. $\square$

From this estimation, we can conclude that

$$s_\infty(\varphi) \geq N - b_j \quad (3.43)$$

and, hence, that
$$\mu(\varphi) \geq N - b_j - 1. \tag{3.44}$$
This criterion, however, provides only lower bounds for the regularity, and one would like to determine the regularity more precisely. As a first step, we will sharpen the estimate (3.38) by exploiting the fact that this relation is true for all j.

3.4.2 The critical exponent of a CQF

We define the 'critical exponent' of the CQF m_0 to be the quantity
$$b = \inf_{j > 0} b_j. \tag{3.45}$$
As remarked before, $b_j \geq 0$, and hence $b \geq 0$. The next result will be useful in the sequel.

Lemma 3.1 *The sequence b_j converges to b.*

Proof. Let $\varepsilon > 0$. There exists an integer j_ε such that $|b_{j_\varepsilon} - b| < \frac{\varepsilon}{2}$. For $j \in \mathbb{N}$, we write $j = n_j j_\varepsilon + r_j$ where $0 \leq r_j < j_\varepsilon$. Then, as before, we have
$$\begin{aligned} B_j &\leq (B_{j_\varepsilon})^{n_j} (B_1)^{r_j} \\ &\leq (B_{j_\varepsilon})^{n_j} (B_1)^{j_\varepsilon}. \end{aligned}$$
By taking the logarithm of both sides and using $j \geq n_j j_\varepsilon$, we deduce that
$$b_j \leq b_{j_\varepsilon} + \frac{b_1}{n_j}. \tag{3.46}$$
If j is sufficiently large, we have $\frac{b_1}{n_j} < \frac{\varepsilon}{2}$, and consequently $b \leq b_j < b + \varepsilon$. This proves that $b = \lim_{j \to \infty} b_j$. $\square$

We can use this lemma to reformulate the result of Proposition 3.3 in terms of b. In particular, we can state that
$$s_\infty(\varphi) \geq N - b. \tag{3.47}$$
More precisely, for all $\varepsilon > 0$, there exists a constant C_ε such that
$$|\hat{\varphi}(\omega)| \leq C_\varepsilon (1 + |\omega|)^{-N+b+\varepsilon} \tag{3.48}$$
for all ω in $\mathbb{R}$. We will see below that this inequality is, in a certain sense, optimal.

3.4.3 The decay of $\hat{\varphi}$

Let $m_0(\omega) = \left(\frac{1+e^{i\omega}}{2}\right)^N p(\omega)$ be a CQF in its factored form. In the rest of this chapter we are going to make the following technical assumption:

$$|p(\pi)| > |p(0)| = 1. \tag{3.49}$$

The reader will note that this is not strictly necessary to demonstrate the results that will follow. One can indeed replace this assumption with weaker, more general hypotheses. However, (3.49) is true in all of the frequently used cases and, in particular, in the case of wavelets with compact support that correspond to filters of minimal length for a fixed number N of vanishing moments. The minimal length CQF of the form (3.33) satisfies the relation

$$|m_{0,N}(\omega)|^2 = \left(\cos^2\left(\frac{\omega}{2}\right)\right)^N P_N\left(\sin^2\left(\frac{\omega}{2}\right)\right) \tag{3.50}$$

where $P_N(y) = \sum_{j=0}^{N-1} \binom{N-1+j}{j} y^j$ is the minimal degree polynomial solution of the equation

$$y^N P_N(1-y) + (1-y)^N P_N(y) = 1. \tag{3.51}$$

In this case we have

$$|p(\pi)|^2 = P_N(1) = \sum_{j=0}^{N-1} \binom{N-1+j}{j} > 1.$$

(The exception is the trivial case of the Haar basis where the situation is entirely clear.)

We now write

$$p_j(\omega) = \prod_{k=1}^{j} p(2^{-k}\omega) \tag{3.52}$$

and let K be the compact set congruent to $[-\pi, \pi]$ that satisfies property (P) of Theorem 2.1.

Lemma 3.2 *There exists a sequence of real numbers $\{\omega_j\}_{j>0}$ such that $2^{-j}\omega_j$ are in K and $p_j(\omega_j) = B_j$. Furthermore, there exist two strictly positive constants C_1 and C_2 such that*

$$C_1 < |2^{-j}\omega_j| < C_2 \quad \text{for all} \quad j > 0. \tag{3.53}$$

Proof. It is clear that p_j is $2^{j+1}\pi$- periodic. Since K is congruent to $[-\pi, \pi]$ modulo 2π, B_j is also the maximum of $|p_j(\omega)|$ on the set $2^j K$.

Now choose ω_j in $2^j K$ so that $|p_j(\omega_j)| = B_j$. We will use the hypothesis (3.49) to demonstrate (3.53).

We note that $|p(\omega+\pi)| > |p(\omega)|$ in a neighborhood of the origin of the form $|\omega| \le C_1$. This can also be written as

$$|2^{-j}\omega| \le C_1 \implies |p(2^{-j}(\omega+2^j\pi))| > |p(2^{-j}\omega)|. \qquad (3.54)$$

Furthermore, we have $p(2^{-j'}(\omega+2^j\pi)) = p(2^{-j'}\omega)$ for $j' < j$. By taking products, we conclude that

$$|2^{-j}\omega| \le C_1 \implies |p_j(\omega+2^j\pi)| > |p_j(\omega)|. \qquad (3.55)$$

But this tells us that the point ω_j must be outside the interval $[-2^j C_1, 2^j C_1]$, which is to say, $C_1 < |2^{-j}\omega_j|$.

Finally, since ω_j is in $2^j K$ and since K is compact, there exists a C_2 such that $|2^{-j}\omega_j| < C_2$, and we have the inequality (3.53). □

Hypothesis (3.49) thus provides us with information about the growth of the sequence ω_j.

Since ω_j is in $2^j K$, we can write (see (2.15))

$$\left|\prod_{k=j+1}^{+\infty} p(2^{-k}\omega_j)\right| \ge \left|\prod_{k=j+1}^{+\infty} m_0(2^{-k}\omega_j)\right| \ge C > 0, \qquad (3.56)$$

where the constant C does not depend on j. Thus from Lemma 3.2 we can conclude that

$$\left|\prod_{k=1}^{+\infty} p(2^{-k}\omega_j)\right| \ge C B_j. \qquad (3.57)$$

Also from Lemma 3.2, $2^j C_1 < |\omega_j| < 2^j C_2$, so that

$$B_j = 2^{jb_j} > C_2^{-b_j}|\omega_j|^{b_j} \ge C_2^{-b_j}|\omega_j|^b.$$

Combined with (3.56), this shows that

$$\left|\prod_{k=1}^{+\infty} p(2^{-k}\omega_j)\right| \ge C'|\omega_j|^b, \qquad (3.58)$$

where the constant C' does not depend on j. This strongly suggests that $|\hat{\varphi}(\omega)|$ cannot decrease at infinity faster than $|\omega|^{-N+b}$. We will see in the next section that this is indeed true.

Observe that if $N-b$ is strictly greater than $\frac{1}{2}$, this decay implies that φ belongs to $L^2(\mathbb{R})$. It turns out, however, that this condition yields an even better result.

Proposition 3.4 *If* $N - b > \frac{1}{2}$, *the hypotheses of Theorem 2.1 are automatically satisfied, and* $\{\varphi(x - k)\}_{k \in \mathbb{Z}}$ *is an orthonormal family.*

Proof. Define

$$\hat{g}_n(\omega) = \prod_{k=1}^{n} m_0(2^{-k}\omega) \, \mathbf{1}_{[-2^n\pi, 2^n\pi]}. \quad (3.59)$$

Following the proof of Theorem 2.1, we know that $\{g_n(x - k)\}_{k \in \mathbb{Z}}$ is an orthonormal system and that the sequence $\hat{g}_n(\omega)$ converges pointwise to $\hat{\varphi}(\omega)$. By assumption, we know that there exists a $j > 0$ such that $N - b_j > \frac{1}{2}$.

Taking absolute values in (3.59), we get

$$\begin{aligned} |\hat{g}_n(\omega)| &= \left| \prod_{k=1}^{n} \left(\frac{1 + e^{i2^{-k}\omega}}{2} \right) \right|^N \left| \prod_{k=1}^{n} p(2^{-k}\omega) \right| \mathbf{1}_{[-2^n\pi, 2^n\pi]} \\ &= \left| \frac{\sin(\omega/2)}{2^n \sin(2^{-n}\omega/2)} \right|^N \left| \prod_{k=1}^{n} p(2^{-k}\omega) \right| \mathbf{1}_{[-2^n\pi, 2^n\pi]}. \end{aligned}$$

For the first term, we have

$$\left| \frac{\sin(\omega/2)}{2^n \sin(2^{-n}\omega/2)} \right| \leq \frac{\pi}{2} \left| \frac{\sin(\omega/2)}{(\omega/2)} \right|$$

for $|\omega| \leq 2^n \pi$. Thus

$$|\hat{g}_n(\omega)| \leq (2\pi)^N (1 + |\omega|)^{-N} \left| \prod_{k=1}^{n} p(2^{-k}\omega) \right| \mathbf{1}_{[-2^n\pi, 2^n\pi]}. \quad (3.60)$$

We estimate the second product using arguments similar to those used in Proposition 3.3. For a fixed ω, we define l to be the unique integer such that $2^l \leq 1 + |\omega| < 2^{l+1}$. Using (3.37), we have the estimate

$$\prod_{k=1}^{n} |p(2^{-k}\omega)| \leq B_m Q_2, \quad (3.61)$$

where $m = \min\{l, n\}$. The same argument that led to (3.41) shows that

$$\prod_{k=1}^{n} |p(2^{-k}\omega)| \leq Q_2 (B_1)^j (1 + |\omega|)^{b_j}. \quad (3.62)$$

Taken together, these estimates show the existence of a constant

C'_j, which depends on j but which is independent of n, such that

$$|\hat{g}_n(\omega)| \leq C'_j(1+|\omega|)^{-N+b_j}$$
$$\leq C'_j(1+|\omega|)^{-\frac{1}{2}-\varepsilon},$$

for some $\varepsilon > 0$. As a consequence of this estimate, we can apply the dominated convergence theorem and conclude that $\hat{g}_n$ tends to $\hat{\varphi}$ in L^2. Knowing that $\hat{\varphi} \in L^2$ yields very simply the conclusions of Theorem 2.1. □

This decay hypothesis is nevertheless too strong to characterize all the multiresolution analyses. One can become convinced of this by a simple example.

In Daubechies (1988) it is shown that the CQFs having exactly four real non-zero coefficients have the parametric representation

$$h_0 = \frac{\beta(\beta-1)}{2(\beta^2+1)}, \qquad h_1 = \frac{1-\beta}{2(\beta^2+1)},$$
$$h_2 = \frac{\beta+1}{2(\beta^2+1)}, \qquad h_3 = \frac{\beta(\beta+1)}{2(\beta^2+1)}.$$

When the real number β is $-1, 1$ or 0, one gets the Haar system. Furthermore, it is possible to show that property (P) is true for all β and that one even has $m_{0,\beta}(\omega) \neq 0$ on $\left[-\frac{\pi}{2}, \frac{\pi}{2}\right]$. On the other hand, when β tends to $+\infty$, the filter approaches the 'degenerate case' $h_0 = h_3 = \frac{1}{2}$ and $h_1 = h_2 = 0$, which corresponds to the Haar system dilated by a factor of 3.

Thus, for each $\varepsilon > 0$, we can find a β large enough so that

$$\left|m_{0,\beta}\left(-\frac{2\pi}{3}\right)\right| = \left|m_{0,\beta}\left(\frac{2\pi}{3}\right)\right| \geq 2^{-\varepsilon}. \tag{3.63}$$

By taking $\omega_j = 2^j \frac{\pi}{3}$, we have

$$|\hat{\varphi}_\beta(\omega_j)| = \prod_{k=0}^{+\infty} \left|m_{0,\beta}\left(2^{-k}\frac{\pi}{3}\right)\right| \prod_{k=1}^{j-1} \left|m_{0,\beta}\left(2^k\frac{\pi}{3}\right)\right|$$
$$= C\left|m_{0,\beta}\left(\frac{2\pi}{3}\right)\right|^{j-1} \simeq C|\omega_j|^{-\varepsilon}.$$

This means that $s_\infty(\varphi_\beta) \leq \varepsilon$. Thus the function $\hat{\varphi}_\beta$ can belong to $L^2(\mathbb{R})$ while, at the same time, it has very poor decay since ε can be taken arbitrarily small. This function is lacunary in the sense that the zones of poor decay are bulges that appear more and more rarely as the frequency increases.

In spite of these disadvantages, the critical exponent allows us to determine the s_p-regularity in such a way that the incurred error will be relatively insignificant for large values of p.

Indeed, in case φ is compactly supported, one has the relation

$$s_\infty(\varphi) - 1 \leq s_1(\varphi) \leq \mu(\varphi) \leq s_\infty(\varphi). \tag{3.64}$$

To check that $\mu(\varphi) \leq s_\infty(\varphi)$, simply notice that if φ is in $\mathcal{C}^\mu$ with $0 < \mu < 1$, then

$$|\hat{\varphi}(\omega)(1 - e^{i\omega t})| \leq \int |\varphi(x) - \varphi(x+t)|\, dx \leq C|t|^\mu.$$

By taking $\omega = \frac{\pi}{t}$, we see that $|\hat{\varphi}(\omega)| \leq \frac{C\pi^\mu}{2}|\omega|^{-\mu}$. For $\mu > 1$, we reach the same conclusion by noting that $|\widehat{\varphi^{(n)}}(\omega)| = |\omega|^n|\hat{\varphi}(\omega)|$. We will now prove a more general estimate for the L^p-Sobolev regularity.

3.5 Estimation of the L^p-Sobolev exponent

This is the result that we will prove.

Theorem 3.1 *Let $m_0(\omega) = \left(\frac{1+e^{i\omega}}{2}\right)^N p(\omega)$ be a CQF that satisfies property (P) and (3.49). Then the function φ generated by the product (3.8) satisfies*

$$N - b - \frac{1}{p} \leq s_p(\varphi) \leq N - b. \tag{3.65}$$

Proof. The proof is technical, but the idea is simple: We show that the integral $\int |\omega|^{\alpha p} |\hat{\varphi}(\omega)|^p\, d\omega$ diverges when α is greater than $N - b$. To do this, we estimate the size and length of the 'bumps' where $|\hat{\varphi}(\omega)|$ is relatively large.

We consider the sequence $\{\omega_j\}_{j>0}$ provided by Lemma 3.2 and introduce the notation, $B = 2^b$.

For each ε in $]0, 1[$, we will construct a sequence of intervals I_j that satisfy the following conditions:

- I_j is in $2^j K$,
- $|p_j(\omega)| \geq \frac{B^j}{2}$ on I_j,
- I_j contains ω_j, and
- the length of I_j is of the order $C(1-\varepsilon)^j$.

It is on these intervals that $|\hat{\varphi}(\omega)|$ will be large enough to cause the integral to diverge.

ESTIMATION OF THE L^P-SOBOLEV EXPONENT

Before launching into the construction of these intervals, we mention that, in the interest of conserving notation, C will denote the generic constant. The value of C will change from place to place, it will usually depend on ε, but C will always be independent of j.

We begin the construction of the I_j by examining the derivative of the function p_j, which is expressed as

$$p'_j(\omega) = \sum_{l=1}^{j} 2^{-l} p'(2^{-l}\omega) \prod_{\substack{k=1 \\ k \neq l}}^{j} p(2^{-k}\omega). \tag{3.66}$$

Consequently

$$\max_{\omega \in \mathbb{R}} |p'_j(\omega)| \leq C \sum_{l=0}^{j-1} 2^{-l} B_l B_{j-l-1}, \tag{3.67}$$

with $C = 2^{-1} \max_{\omega \in \mathbb{R}} |p'(\omega)|$. Using Lemma 3.1, we know that there exists an l_ε such that, for all $l \geq l_\varepsilon$,

$$B_l \leq (B+\varepsilon)^l. \tag{3.68}$$

Since there are only a finite number of $l < l_\varepsilon$, there is a C, which depends on ε but not on j, such that

$$\max_{\omega \in \mathbb{R}} |p'_j(\omega)| \leq C(B+\varepsilon)^j. \tag{3.69}$$

We know that $|p_j(\omega_j)| = B_j \geq B^j$. Thus, $|p_j(\omega)| \geq B^j/2$ in some interval containing ω_j. In fact, the estimate (3.69) implies that $|p_j(\omega)| \geq B^j/2$ whenever $|\omega_j - \omega| \leq C'(1-\varepsilon)^j$ with $C' = 1/(2C)$, where C is the constant in (3.69).

Now let J_j be the intervals defined by $|\omega_j - \omega| \leq C'(1-\varepsilon)^j$. These intervals satisfy all of the properties we need except the first one, namely, we are not assured that J_j is in $2^j K$. To fix this, define I_j to be that interval in the intersection of J_j and $2^j K$ that contains ω_j. Here we us the fact that $K = \cup_{i=1}^{n} K_i$, where the K_i are disjoint closed intervals.

As a last step in this construction, we must be sure that the intervals I_j are long enough, that is, we want $|I_j| \geq C(1-\varepsilon)^j$ for some constant C. Let $k = \min_{1 \leq i \leq n} |K_i|$. Then there exists a j_ε such that

$$|2^{-j} J_j| = 2C' \left(\frac{1-\varepsilon}{2}\right)^j \leq k$$

whenever $j \geq j_\varepsilon$. For these j,
$$|2^{-j} I_j| \geq C'\left(\frac{1-\varepsilon}{2}\right)^j$$
because at least half of the interval $2^{-j} J_j$ intersects one of the K_i to form $2^{-j} I_j$. For the $j < j_\varepsilon$ we choose a finite number of constants c_j so that
$$|2^{-j} I_j| \geq c_j \left(\frac{1-\varepsilon}{2}\right)^j.$$
By taking $C = \min(\pi, C', c_1, \ldots c_{j_\varepsilon - 1})$, we are guaranteed that $|I_j| \geq C(1-\varepsilon)^j$ for all j. For the record, we note that
$$C(1-\varepsilon)^j \leq |I_j| \leq |J_j| = 2C'(1-\varepsilon)^j \qquad (3.70)$$
and that $C(1-\varepsilon)^j < \pi$, which will be handy in a moment.

The last step before doing the integration is to show that there are plenty of intervals I_j that do not intersect. For this, choose a positive integer a so that $2^a C_1 - C_2 > C'$, where C_1 and C_2 are the constants in Lemma 3.2 and C' is the constant in the definition of J_j. A margin computation shows that
$$2^{aj} C_2 + C'(1-\varepsilon)^j < 2^{a(j+1)} C_1 - C'(1-\varepsilon)^{j+1},$$
which means that the intervals J_{aj}, and consequently the I_{aj}, are disjoint for all $j > 0$.

To finish the proof of the theorem, we will study the integral
$$\int_{\cup I_{aj}} |\omega|^{\alpha p} |\hat{\varphi}(\omega)|^p \, d\omega, \qquad (3.71)$$
which, by the last remark, reduces to looking at the series
$$\sum_{j>0} \int_{I_{aj}} |\omega|^{\alpha p} |\hat{\varphi}(\omega)|^p \, d\omega. \qquad (3.72)$$
Begin by decomposing $\hat{\varphi}(\omega)$ as follows:
$$\hat{\varphi}(\omega) = p_{aj}(\omega) \prod_{k=1}^{aj} \left(\frac{1 + e^{i 2^{-k} \omega}}{2}\right)^N \prod_{k=aj+1}^{+\infty} m_0(2^{-k} \omega). \qquad (3.73)$$
We will deal with each of the three factors separately.
- We have already shown that $|p_{aj}(\omega)| \geq \frac{B^{aj}}{2}$ on I_{aj}.

ESTIMATION OF THE L^p-SOBOLEV EXPONENT

- Since I_{aj} is in $2^{aj}K$, we know from the proof of Theorem 2.1 that there exists a constant C, which is independent of j, such that

$$\left|\prod_{k=aj+1}^{+\infty} m_0(2^{-k}\omega)\right| \geq C > 0 \quad \text{for all } \omega \text{ in } I_{aj}. \tag{3.74}$$

- Finally, we have

$$\left|\prod_{k=1}^{aj}\left(\frac{1+e^{i2^{-k}\omega}}{2}\right)^N\right| = \left|\frac{\sin(\frac{\omega}{2})}{2^{aj}\sin(2^{-aj}(\frac{\omega}{2}))}\right|^N \geq \left|\frac{2\sin(\frac{\omega}{2})}{\omega}\right|^N \tag{3.75}$$

for all ω.

If $\omega \in I_{aj}$, then its distance from the origin is greater than $2^{aj}C_1 - C'$ and

$$\int_{I_{aj}} |\omega|^{\alpha p} |\hat{\varphi}(\omega)|^p \, d\omega \geq 2^{-a}(2^a C_1 - C')\, 2^{\alpha p a j} \int_{I_{aj}} |\hat{\varphi}(\omega)|^p \, d\omega, \tag{3.76}$$

where $2^a C_1 - C' > 0$ because $2^a C_1 - C_2 > C'$.

From (3.74) and (3.75) we have

$$\int_{I_{aj}} |\hat{\varphi}(\omega)|^p \, d\omega \geq CB^{paj} \int_{I_{aj}} \left|\frac{2\sin(\frac{\omega}{2})}{\omega}\right|^{Np} d\omega. \tag{3.77}$$

If $\omega \in I_{aj}$, then $|\omega| \leq 2^{aj}C_2 + C'$ and

$$\int_{I_{aj}} \left|\frac{2\sin(\frac{\omega}{2})}{\omega}\right|^{Np} d\omega \geq 2^{Np}(C_2+C')^{-Np} 2^{-ajNp} \int_{I_{aj}} \left|\sin\left(\frac{\omega}{2}\right)\right|^{Np} d\omega. \tag{3.78}$$

The last integral is estimated using the following three inequalities: $I_{aj} \geq C(1-\varepsilon)^{aj}$, $C(1-\varepsilon)^{aj} \leq \pi$, and $\sin(\frac{\omega}{2}) \geq \frac{\omega}{4}$ for $0 \leq \omega \leq \pi$. We then have

$$\int_{I_{aj}} \left|\sin\left(\frac{\omega}{2}\right)\right|^{Np} d\omega \geq \int_0^{C(1-\varepsilon)^{aj}} \left(\frac{\omega}{4}\right)^{Np} d\omega$$

$$= 4^{-Np} \frac{C^{Np+1}}{Np+1}(1-\varepsilon)^{aj(Np+1)}$$

Combining the last four inequalities shows that

$$\int_{I_{aj}} |\omega|^{\alpha p}|\hat{\varphi}(\omega)|^p \, d\omega \geq C\, 2^{ajp(\alpha-N+b)}(1-\varepsilon)^{aj(Np+1)} \tag{3.79}$$

for some constant C.

If $\alpha > N - b$, $2^{p(\alpha-N+b)} > 1$, and we can choose ε so that $2^{p(\alpha-N+b)}(1-\varepsilon)^{(Np+1)} > 1$. Thus

$$\int_{I_{\alpha j}} |\omega|^{\alpha p} |\hat{\varphi}(\omega)|^p \, d\omega \geq Cr^j, \tag{3.80}$$

where $r > 1$, and the series (3.72) diverges. From this we conclude that the integral $\int |\omega|^{\alpha p} |\hat{\varphi}(\omega)|^p \, d\omega$ diverges when α is strictly greater than $N - b$, and this proves the theorem.

One should note that, in our argument, the choice of ε to ensure that $r > 1$ does not depend on C or a. On the other hand, both C and a depend on ε. □

Theorem 3.1 implies what was suggested by (3.58), namely, that $s_\infty = N - b$. Specifically, (3.65), $s_p + \frac{1}{p} \geq s_\infty$, and (3.47) imply that

$$N - b + \frac{1}{p} \geq s_p + \frac{1}{p} \geq s_\infty \geq N - b.$$

Letting $p \to +\infty$ shows that $s_\infty = N - b$ and that $s_p + \frac{1}{p} \downarrow s_\infty$.

3.6 Applications

3.6.1 Estimating the critical exponent

It is desirable to be able to estimate b using reasonable computations. By definition, we always have $b \leq b_j$ for all j in $\mathbb{N}^*$, and this gives upper bounds for the critical exponent. To obtain very sharp upper bounds for b, it often pays to introduce another sequence $d_j = \frac{1}{j \log 2} \log D_j$, where D_j is defined by

$$D_j = \sup_{\omega \in \mathbb{R}} \min_{l \in \{1,\ldots,j\}} \left| \prod_{k=1}^{l} p(2^{-k}\omega) \right|^{j/l}. \tag{3.81}$$

By writing

$$d_j = \frac{1}{\log 2} \sup_{\omega \in \mathbb{R}} \min_{l \in \{1,\ldots,j\}} \log \left| \prod_{k=1}^{l} p(2^{-k}\omega) \right|^{1/l}, \tag{3.82}$$

it is clear that d_j is a decreasing sequence and that $d_j \leq b_j$. In fact, this sequence also converges to b. It may, however, converge faster than b_j. To check the convergence, we only need to show that

$$B_n \leq (D_j)^{n/j}(B_1)^j \tag{3.83}$$

for all $n \geq 1$. If this holds, we have
$$b_n \leq d_j + \frac{jB_j}{n\log 2}$$
by taking the logarithm of both sides and dividing by $\log 2$. Fixing j and letting n tend to $+\infty$ shows that $b \leq d_j$. Thus we have $b \leq d_j \leq b_j$, which proves that d_j converges to b.

We prove (3.83) by induction on n. First note that
$$B_n \leq (B_1)^n \leq (D_j)^{n/j}(B_1)^j$$
when $n \leq j$, since D_j and B_1 are greater than or equal to 1. For $n > j$, we fix $\omega \in \mathbb{R}$ and define $l_\omega \in \{1, \ldots, j\}$ so that
$$\min_{l \in \{1,\ldots,j\}} \left|\prod_{k=1}^{l} p(2^{-k}\omega)\right|^{j/l} = \left|\prod_{k=1}^{l_\omega} p(2^{-k}\omega)\right|^{j/l_\omega}.$$
We thus have
$$\begin{aligned}\left|\prod_{k=1}^{n} p(2^{-k}\omega)\right| &\leq (D_j)^{l_\omega/j} B_{n-l_\omega} \\ &\leq (D_j)^{l_\omega/j}(D_j)^{(n-l_\omega)/j}(B_1)^j \\ &= (D_j)^{n/j}(B_1)^j.\end{aligned}$$
Here we use the induction hypothesis to get the second inequality. Since this is true for all ω, (3.83) follows.

We have two different techniques to develop lower bounds. The first method uses ergodic theory. Indeed, for almost all ω in $[-\pi, \pi]$, we have
$$\begin{aligned}\lim_{j \to +\infty} & \left[\frac{1}{j\log 2} \log\left(\prod_{k=0}^{j} |p(2^k\omega)|\right)\right] \\ &= \lim_{j \to +\infty} \frac{1}{j\log 2} \sum_{k=0}^{j} \log |p(2^k\omega)| \\ &= \frac{1}{2\pi \log 2} \int_0^{2\pi} \log |p(\omega)|\, d\omega,\end{aligned}$$
and consequently
$$b \geq \frac{1}{2\pi \log 2} \int_0^{2\pi} \log |p(\omega)|\, d\omega. \tag{3.84}$$
The second method consists in looking at the values of p at points

that 'escape' the ergodic theory for the transformation $\omega \mapsto 2\omega$ modulo 2π. The simplest of these are clearly the finite cycles that we introduced in Chapter 2.

Let $\{\delta_1, \ldots, \delta_n\}$ be one of these sets. We can then write, for all $j > 0$,

$$\prod_{k=1}^{jn} p(2^k \delta_1) = \left(\prod_{k=1}^{n} p(\delta_k)\right)^j. \tag{3.85}$$

It follows from this that

$$b \geq \frac{1}{n \log 2} \log\left(\left|\prod_{k=1}^{n} p(\delta_k)\right|\right). \tag{3.86}$$

Among others, we have, by considering the simplest cycle, the estimate

$$b \geq \frac{1}{2 \log 2} \log\left(\left|p\left(\frac{2\pi}{3}\right) p\left(-\frac{2\pi}{3}\right)\right|\right). \tag{3.87}$$

In practice, one notices that the second technique is more effective than the first for estimating the critical exponent accurately. More precisely, in the case of wavelets with compact support corresponding to filters given by (3.50), the numerical calculation of values of b_j shows that the maximum of the function

$$|p_{N,j}(2^j \omega)| = \left|\prod_{k=0}^{j-1} p_N(2^k \omega)\right| \tag{3.88}$$

tends to occur at $\omega = \pm \frac{2\pi}{3}$ for large j. In this case, one can actually prove that b_j converges to

$$\frac{1}{2 \log 2} \log\left(\left|p_N\left(\frac{2\pi}{3}\right) p_N\left(-\frac{2\pi}{3}\right)\right|\right) = \frac{\log\left(\left|p_N\left(\frac{2\pi}{3}\right)\right|\right)}{\log 2}. \tag{3.89}$$

Proposition 3.5 *The critical exponent of a CQF of type* (3.50) *is given by*

$$b(N) = \frac{1}{\log 2} \log\left(\left|p_N\left(\frac{2\pi}{3}\right)\right|\right). \tag{3.90}$$

Proof. The proof of this result was first given in Cohen and Conze (1992). It is based on an estimate that we state as a lemma.

APPLICATIONS 91

Lemma 3.3 *For all $N \geq 1$, the functions p_N satisfy*

$$|p_N(\omega)| \leq |p_N(\frac{2\pi}{3})| \quad \text{for all} \quad |\omega| \leq \frac{2\pi}{3}, \qquad (3.91)$$

and

$$|p_N(\omega)p_N(2\omega)| \leq |p_N(\frac{2\pi}{3})|^2 \quad \text{for all} \quad \frac{2\pi}{3} \leq |\omega| \leq \pi. \qquad (3.92)$$

Proof. Recall that $|p_N(\omega)|^2 = P_N(\sin^2 \frac{\omega}{2})$, where

$$P_N(y) = \sum_{j=0}^{N-1} \binom{N-1+j}{j} y^j.$$

Since P_N has positive coefficients, it is clear that (3.91) is true. The second relation, (3.92), is equivalent to

$$P_N(y)P_N(4y(1-y)) \leq [P_N(\frac{3}{4})]^2 \quad \text{for} \quad y \in [\frac{3}{4}, 1]. \qquad (3.93)$$

We use induction on the degree of the polynomial to prove (3.93).

Thus, for the induction hypothesis, assume that R_{n-1} is a polynomial of degree $n-1$ with positive coefficients that satisfies (3.93). If $R_n(x) = R_{n-1}(x) + a_n x^n$, then we have

$$\begin{aligned}R_n(y)R_n(4y(1-y)) &= R_{n-1}(y)R_{n-1}(4y(1-y)) \\ &+ a_n^2(y4y(1-y))^n + a_n(4y(1-y))^n R_{n-1}(y) \\ &+ a_n y^n R_{n-1}(4y(1-y)).\end{aligned}$$

By the induction hypothesis, the first term on the right-hand side is dominated on $[\frac{3}{4}, 1]$ by its value at $y = \frac{3}{4}$. It is easy to show that the same property holds for the second term, and it is only slightly more complicated to show that it holds for the third term. In the latter case, one shows that the relation holds term by term, that is, that $(4y(1-y))^n y^k \leq (\frac{3}{4})^n (\frac{3}{4})^k$. Finally, we must show that

$$y^n R_{n-1}(4y(1-y)) \leq (\frac{3}{4})^n R_{n-1}(\frac{3}{4}) \quad \text{for} \quad y \in [\frac{3}{4}, 1]. \qquad (3.94)$$

This condition will hold if the coefficients of R_n grow fast enough. More precisely, (3.94) will be true if

$$\frac{a_k}{a_{k-1}} \geq \frac{2}{3}\frac{n}{k}, \quad k = 1, 2, \ldots, n-2, \text{ and } \frac{a_{n-1}}{a_{n-2}} \geq \frac{4}{3}\frac{n}{n-2}. \qquad (3.95)$$

To check this, write $f(y) = y^n R_{n-1}(4y(1-y))$. Then

$$f'(y) = y^{n-1}[nR_{n-1}(4y(1-y)) + 4y(1-2y)R'_{n-1}(4y(1-y))].$$

Since $4y(1-2y)$ is negative and decreasing on $[\frac{3}{4}, 1]$,

$$f'(y) \leq y^{n-1}[nR_{n-1}(4y(1-y)) - \frac{3}{2}R'_{n-1}(4y(1-y))].$$

Condition (3.95) ensures that the right-hand side of this inequality is less than or equal to zero for $y \in [\frac{3}{4}, 1]$, and this, in turn, implies (3.94).

We return to the specific case of the sequence P_N. One can check numerically that (3.93) holds if $N < 7$. For $N \geq 7$, the conditions (3.95) are satisfied by the coefficients a_n of P_N when $3 \leq n \leq N-1$. Finally, the induction is initialized by checking directly that (3.93) is true for $\sum_{n=0}^{2} a_n x^n$ when $N > 7$. This concludes the proof of the lemma. □

The proof of the proposition follows directly from the lemma. By definition,

$$d_2(N) = \frac{1}{2\log 2} \log \left(\sup_{\omega \in \mathbb{R}} \min \{ |p_N(\omega)|^2, |p_N(\omega)p_N(2\omega)| \} \right).$$

By the lemma, the right-hand side is dominated by $\frac{\log |p_N(\frac{2\pi}{3})|}{\log 2}$, which implies that

$$d_2(N) \leq \frac{\log |p_N(\frac{2\pi}{3})|}{\log 2}. \tag{3.96}$$

Since $b(N) \leq d_2(N)$, (3.96) and (3.87) prove the proposition. □

REMARKS

Equality (3.90) has no chance to be true for all wavelets with compact support. Formula (3.50) is a particular case of the general formula

$$|m_0(\omega)|^2 = (\cos^2(\frac{\omega}{2}))^N \left[P_N(\sin^2(\frac{\omega}{2})) + R(\frac{1}{2} - \sin^2(\frac{\omega}{2})) \right] \tag{3.97}$$

where R is an odd polynomial such that the right-hand side is positive. (One then uses a lemma of Riesz to recover m_0.) This polynomial introduces too many degrees of freedom for (3.50) to be true. One could impose the condition that $m_0(\pm\frac{2\pi}{3}) = 0$, for example.

The general formula (3.97) is related to several open problems. In particular, what is the optimal degree of regularity that one can obtain when one fixes N or the number of coefficients of the filter?

3.6.2 The behavior for large values of N

We now return to the formula (3.50) and present several results that arise when N tends to $+\infty$.

Proposition 3.6 *The filter $m_{0,N}(\omega)$ tends toward a perfect band-pass filter for the interval $\left[-\frac{\pi}{2}, \frac{\pi}{2}\right]$. More precisely, one has*

$$\lim_{N \to +\infty} |m_{0,N}(\omega)| = 1 \quad \text{if } \omega \in \left]-\frac{\pi}{2}, \frac{\pi}{2}\right[, \tag{3.98}$$

$$\lim_{N \to +\infty} |m_{0,N}(\omega)| = 0 \quad \text{if } \omega \in \left[-\pi, -\frac{\pi}{2}\right] \cup \left[\frac{\pi}{2}, \pi\right], \tag{3.99}$$

and, given $\varepsilon > 0$, the convergence is uniform on $\left]-\frac{\pi}{2}+\varepsilon, \frac{\pi}{2}-\varepsilon\right[$ and $\left[-\pi, -\frac{\pi}{2}-\varepsilon\right] \cup \left[\frac{\pi}{2}+\varepsilon, \pi\right]$.

Proof. It is clearly sufficient to prove (3.99), and for this we analyse

$$|m_{0,N}(\omega)|^2 = (1-y)^N P_N(y)$$

when $y \in \left]\frac{1}{2}, 1\right]$ and where $y = \sin^2(\frac{\omega}{2})$. For these values we have

$$\begin{aligned} P_N(y) &= \sum_{j=0}^{N-1} \binom{N-1+j}{j} y^j \\ &= \sum_{j=0}^{N-1} \binom{N-1+j}{j} 2^{-j} (2y)^j \\ &\leq (2y)^{N-1} P_N(\tfrac{1}{2}). \end{aligned}$$

By equation (3.51) that defines P_N, we have $2^{-N+1} P_N(\frac{1}{2}) = 1$, and thus $P_N(\frac{1}{2}) = 2^{N-1}$. (Alternatively, one can manipulate binomial coefficients and see that

$$(1-y) P_{N+1}(y) = P_N(y) + \binom{2N-1}{N} y^N (1-2y).$$

Then $P_N(\frac{1}{2}) = 2^{N-1}$ follows immediately by induction.) However it is done, this shows that for $y \in \left]\frac{1}{2}, 1\right]$, $P_N(y) \leq (4y)^{N-1} \leq (4y)^N$. Thus for all $\omega \in \left[-\pi, -\frac{\pi}{2}\right[\cup \left]\frac{\pi}{2}, \pi\right]$, we have

$$|m_{0,N}(\omega)|^2 \leq \left[\cos^2(\tfrac{\omega}{2}) [4\sin^2(\tfrac{\omega}{2})]\right]^N = [\sin^2(\omega)]^N. \tag{3.100}$$

The proposition follows directly from this estimate. $\square$

This result shows that the ability of CQFs of type (3.50) to separate frequencies in and out of $\left]-\frac{\pi}{2}, \frac{\pi}{2}\right[$ increases with the degree of cancellation N, that is, with the order of the zero of m_0 at π.

94   THE REGULARITY OF SCALING FUNCTIONS AND WAVELETS

Note that the argument here has involved only the modulus of the transfer function. We do not know if it is possible to choose solutions of (3.50), using the Riesz lemma, so that one is assured that $m_{0,N}$ converges to the ideal band-pass filter whose transfer function is 1 on $[-\frac{\pi}{2}, \frac{\pi}{2}]$ and 0 elsewhere.

We can now investigate the asymptotic behavior of the critical exponent $b(N)$ of $m_{0,N}$. For this, we begin with the relation

$$\left| p_N \left(\frac{2\pi}{3} \right) \right|^2 = \sum_{j=0}^{N-1} \binom{N-1+j}{j} \left(\frac{3}{4} \right)^j$$

$$\geq \left(\frac{3}{4} \right)^{N-1} \sum_{j=0}^{N-1} \binom{N-1+j}{j}$$

$$= \left(\frac{3}{4} \right)^{N-1} \frac{1}{2} \binom{2N}{N}.$$

Here we have used the identity

$$\sum_{j=0}^{N-1} \binom{N-1+j}{j} = \binom{2N-1}{N} = \frac{1}{2} \binom{2N}{N}.$$

One can easily check by induction that $\binom{2N}{N} \geq \frac{1}{2} \frac{4^N}{\sqrt{N}}$. Thus

$$\frac{3^{N-1}}{\sqrt{N}} \leq \left| p_N \left(\frac{2\pi}{3} \right) \right|^2.$$

By the same argument that we used in Proposition 3.5, we deduce that

$$\left| p_N \left(\frac{2\pi}{3} \right) \right|^2 \leq 3^{N-1}.$$

Combining these inequalities shows that

$$\frac{3^{N-1}}{\sqrt{N}} \leq \left| p_N \left(\frac{2\pi}{3} \right) \right|^2 \leq 3^{N-1}. \tag{3.101}$$

This relation implies that the logarithm of $\left| p_N \left(\frac{2\pi}{3} \right) \right|^2$ is asymptotic to $N \log 3$. From this we conclude that

$$\lim_{N \to +\infty} \left(\frac{1}{N}(N - b(N)) \right) = 1 - \frac{\log 3}{2 \log 2} \approx 0.2075. \tag{3.102}$$

These asymptotic estimates argue for our spectral approach to the study of regularity. It is true that the error in the estimates of the Hölder exponent given by (3.65) can be avoided by calculations

APPLICATIONS

made in the spatial domain. Indeed, by a matrix study, Daubechies and Lagarias (1991, 1992) obtained the exact local Hölder exponents of the function φ. However, this method becomes very complex when N takes large values. Another method, based on the transition operator that was introduced in Chapter 2, can also be used to characterize the Sobolev exponent $s_2(\varphi)$. This technique will be used in the next chapter. It also becomes complex for large values of N.

On the other hand, our estimates become more interesting in the case of large N because the relative error is constant. Thus we have exactly

$$\lim_{N \to +\infty} \frac{\alpha_N}{N} = 1 - \frac{\log 3}{2 \log 2} \approx 0.2075. \qquad (3.103)$$

Figure 3.5 exhibits the graph of $r(N) = s_\infty(\varphi_N) = N - b(N)$. One of the things that all of this tells us is that the regularity of φ is not related with 'slope 1' to the number of vanishing moments of the wavelets. In fact, for a regularity of class $\mathcal{C}^r$, it is necessary to have about $N \approx 5r$ zero moments. The corresponding filter would then have about $2N \approx 10r$ coefficients.

It is important to keep this estimate in mind when making a compromise between the regularity that one wants for the wavelets and the size of the calculations implied by the length of the CQF that one uses.

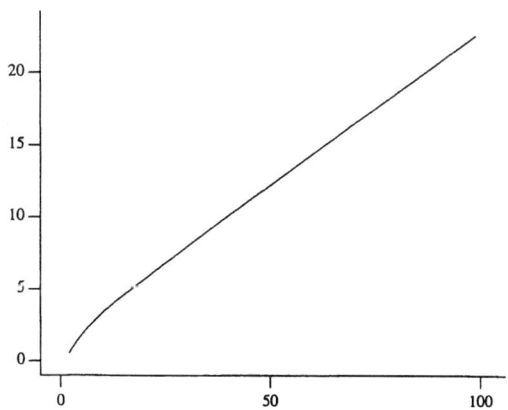

Figure 3.5 *Graph of $r(N) = s_\infty(\varphi_N) = N - b(N)$.*

CHAPTER 4

Biorthogonal wavelet bases

4.1 Introduction

In this chapter we will extend the concepts and methods developed in the preceding chapters to introduce biorthogonal wavelet bases. We begin in the finite dimensional setting to review the principles of biorthogonality.

Given an arbitrary basis $\{e_j\}_{j=1,...,n}$, there exists a unique dual basis $\{\tilde{e}_j\}_{j=1,...,n}$ such that

$$\langle e_i, \tilde{e}_j \rangle = \delta_{i,j}. \tag{4.1}$$

Each vector x can then be decomposed as

$$x = \sum_{j=1}^{n} \langle x, e_j \rangle \tilde{e}_j \tag{4.2}$$

or as

$$x = \sum_{j=1}^{n} \langle x, \tilde{e}_j \rangle e_j. \tag{4.3}$$

We see that one of the bases is used to analyze the element x (by taking scalar products) and the other basis is used for the synthesis (by taking linear combinations). Of course, orthogonality corresponds to the particular case where the two families $\{e_j\}$ and $\{\tilde{e}_j\}$ are identical.

Such dual systems have been well known and have been used for a long time, for example, in harmonic analysis. On the other hand, the existence of biorthogonal bases for $L^2(\mathbb{R})$ that have a multiscale structure is a relatively new discovery. The first example was given in work by Philippe Tchamitchian (1987) and constitutes a particular case of the constructions that we will present. These constructions are also described in Cohen, Daubechies, and Feauveau (1992).

But why is one interested in such a generalization of multiscale bases for $L^2(\mathbb{R})$? In the first place, as we will see, the multiscale decompositions of the type

$$f(x) = \sum_{j \in \mathbb{Z}} 2^j \sum_{k \in \mathbb{Z}} \langle f, \tilde{\psi}(2^j x - k) \rangle \psi(2^j x - k) \qquad (4.4)$$

rely on pyramid algorithms of the same type as those that have been described in Chapter 1. The innovation in the biorthogonal case comes from the considerable flexibility that one has for the choice of filters. This allows one to avoid two of the principal constraints associated with the use of CQFs with finite impulse response, namely, asymmetry and the numerical complexity of their coefficients.

We will see, for example, that one can use the filter whose transfer function is given by $m_0(\omega) = \left(\frac{1+e^{i\omega}}{2}\right)^N$, which corresponds to multiscale approximation in the space of spline functions of order N. In the framework of orthonormal wavelets, the filters associated with spline functions have an infinite impulse response. Thus, except for the Haar system, these approximation spaces (in spite of their attractive simplicity) were until now unattractive for exact reconstruction using orthonormal pyramid algorithms.

In extending the previously established results, our point of departure will be the digital analysis of discrete signals in $l^2(\mathbb{Z})$. With this orientation, we wish to develop, by breaking away from the restrictions imposed by the CQFs, the functional analytic tools that generalize the orthonormal wavelet bases for $L^2(\mathbb{R})$. We will thus begin by reviewing the principles of subband coding that have appeared frequently in the signal processing literature (Esteban and Galand (1977), Smith and Barnwell (1986), Vaidyanathan (1992)).

Formula (4.4) tells us that we will have an analyzing wavelet $\tilde{\psi}$ and a reconstruction wavelet ψ. In the examples that we present in Section 4.5, we will see that the properties useful for analysis (oscillations, vanishing moments) can be allocated to the function $\tilde{\psi}$ while the properties attractive for synthesis (regularity, simplicity) can be assigned to the function ψ. The separation of these two tasks, which is related to the flexibility one has in the biorthogonal case, ought to prove useful in many applications.

We conclude this introduction by indicating the problem that lies at the heart of the construction of biorthogonal wavelet bases.

Formula (4.4) exhibits a decomposition into the subspaces W_j that are generated by the functions $\psi(2^j x - k)$. Although we have $L^2(\mathbb{R}) = \bigoplus_{j \in \mathbb{Z}} W_j$, this is not an orthogonal sum. It is easy to verify that the 'angle' between two consecutive spaces W_j and W_{j+1} does not depend on j. However, one might reasonably fear that 'stacking up' these non-orthogonal subspaces could fail to produce unconditional bases for $L^2(\mathbb{R})$. This point is quite crucial, for it is related to the stability of the pyramid algorithms associated with the biorthogonal wavelets.

The central theorem of Section 4.3 responds to this issue and provides sufficient conditions, in a very general setting, for the construction of unconditional bases for $L^2(\mathbb{R})$. The conditions given in this section involve the critical exponents of the filters and are thus 'spectral' conditions. These conditions are reasonable in that they can be used to construct 'realistic' bases for various applications.

We take a different point of view in Section 4.4. There we will use the transition operator P_0 that was defined in Chapter 2 to establish necessary and sufficient conditions on the filters so that they generate stable (i.e., unconditional) biorthogonal bases. These conditions will involve the coefficients of the filters more directly than the spectral conditions of Section 4.3.

4.2 General principles of subband coding

Subband coding is, as will be shown, a simple operation that consists of a sequence of linear transformations on a digital signal. For practical applications, the attractiveness of this decomposition will depend on the statistical nature of the signals that one processes. As we will see in the next chapter, subband coding can, in certain cases, be very effective for signal compression when it is used in conjunction with non-linear operations such as setting a threshold and quantization.

The signals of interest are given as sequences $\{s_k\}$ that belong to $l^2(\mathbb{Z})$. As usual, we will write the discrete Fourier transform as

$$S(\omega) = \sum s_k e^{-ik\omega}. \qquad (4.5)$$

S is then in $L^2[0, 2\pi]$. The action of a discrete filter $\{f_k\}$ on a signal $\{s_k\}$ is expressed in the spectral domain as multiplication of S by the transfer function F, which is defined by $F(\omega) = \sum f_k e^{-ik\omega}$.

Note that, in all that follows, we assume that the filters have finite impulse response. From all evidence, these filters are the most realistic from the point of view of technological applications. As a consequence of this assumption, the transfer functions F will be trigonometric polynomials. However, all the results that we prove in this chapter can be generalized in some way to 'reasonable' infinite impulse response filters such as, for example, filters with rational transfer functions.

Subband coding consists in first applying two filters H_1 and H_2 to the signal. The filter H_1 is assumed to be low-pass and H_2 is assumed to be high-pass, in the sense that we require

$$H_1(0) = H_2(\pi) = 1 \quad \text{and} \quad H_1(\pi) = H_2(0) = 0. \tag{4.6}$$

After filtering, the two signals are decimated (every other value is dropped) so the two signals together contain the same amount of information as the original signal.

One can then begin a pyramid decomposition by iterating this process on the signal filtered by H_1, just as in the algorithm described in Chapter 1.

It is necessary, however, to be assured that at each step one can reconstruct the signal exactly. For this, one first inserts zeros in place of the values that were dropped and then one applies interpolating filters K_1 and K_2 to each of the signals. Finally, the outputs of these two filters are added to recapture the original signal.

Expressed in the frequency domain, these operations are as follows: We start with the function S.

a) The filtered signals are given by $S_1(\omega) = H_1(\omega)S(\omega)$ and $S_2(\omega) = H_2(\omega)S(\omega)$.

b) After decimation and insertion of zeros we have

$$S_1^d(\omega) = H_1(\omega)S(\omega) + H_1(\omega + \pi)S(\omega + \pi),$$

$$S_2^d(\omega) = H_2(\omega)S(\omega) - H_2(\omega + \pi)S(\omega + \pi).$$

By convention we omit the factor $\frac{1}{2}$ and we decimate the odd points in S_1 and the even points in S_2.

c) After interpolation and adding the results we have

$$S_r(\omega) = K_1(\omega)S_1^d(\omega) + K_2(\omega)S_2^d(\omega),$$

which is

$$\begin{aligned}S_r(\omega) &= S(\omega)\left[K_1(\omega)H_1(\omega) + K_2(\omega)H_2(\omega)\right] \\ &+ S(\omega + \pi)[K_1(\omega)H_1(\omega + \pi) - K_2(\omega)H_2(\omega + \pi)].\end{aligned}$$

The condition for exact reconstruction is thus given by the system

$$\begin{cases} H_1(\omega)K_1(\omega) + H_2(\omega)K_2(\omega) = 1, \\ H_1(\omega+\pi)K_1(\omega) - H_2(\omega+\pi)K_2(\omega) = 0. \end{cases} \quad (4.7)$$

If H_1 and H_2 are fixed, K_1 and K_2 are the solutions of this system.

However, if we wish to have solutions that are filters with finite impulse response, we must avoid dividing by a non-constant trigonometric polynomial. Thus we choose

$$K_1(\omega) = H_2(\omega+\pi) \quad \text{and} \quad K_2(\omega) = H_1(\omega+\pi). \quad (4.8)$$

This is the pair of solutions we get when the determinant is required to be -1, that is,

$$H_1(\omega)H_2(\omega+\pi) + H_1(\omega+\pi)H_2(\omega) = 1. \quad (4.9)$$

Relation (4.9) ensures exact reconstruction of the signal, at least theoretically. To find H_1 and H_2 that satisfy (4.9), we can begin by looking for other constraints.

In 1977, Esteban and Galand proposed having

$$H_2(\omega) = H_1(\omega+\pi), \quad (4.10)$$

which gives

$$(H_1(\omega))^2 + (H_1(\omega+\pi))^2 = 1. \quad (4.11)$$

Unfortunately, (4.11) can only be approximated by filters with finite impulse response, that is, (4.11) has no exact solutions in finite trigonometric polynomials.

In 1983, Smith and Barnwell constructed the solution that corresponds to orthonormal wavelets, which is

$$H_2(\omega) = \overline{H_1(\omega+\pi)}. \quad (4.12)$$

In this case, we recover the familiar identity

$$|H_1(\omega)|^2 + |H_1(\omega+\pi)|^2 = 1. \quad (4.13)$$

It is then possible to find finite impulse response filters that satisfy (4.13). Ingrid Daubechies gave a complete classification of these filters in Daubechies (1988).

Nevertheless, the CQFs provide a rather restrictive framework, and this happens for two principal reasons:

• Their coefficients are not numerically simple. They are algebraic numbers that are difficult to compute explicitly for long filters.

• These filters (and hence their associated compactly supported wavelets) cannot have any symmetry, except for the Haar system (see Daubechies (1988)). This is a disadvantage for image processing where one often requires real filters with linear phase if one is to avoid certain artifacts.

To get around these difficulties, it is necessary to abandon orthogonality. But this is not done lightly, for it is the orthogonality that gives us control over the quadratic norms in the decomposition and reconstruction. Returning to (4.9) in all its generality, we see that if the coefficients of H_1 are fixed, it is possible to calculate the coefficients of H_2 by solving a classical linear system. Moreover, we note that this system does not have a unique solution.

Pairs of solutions have been introduced by Martin Vetterli (1986). In this work, the numerical expressions of the coefficients are remarkably simple. In fact, one can choose the coefficients so that they all have a finite binary expansion. This is clearly a very useful property for the implementation of the algorithms.

Furthermore, in this situation it is possible to choose one of the filters so that it is well matched to the type of processing one wishes to do. This implies a flexibility that is very promising for applications. We will see, for example, that the class of filter introduced by Burt and Adelson (1983) fits perfectly with this theory.

In view of the results that have been developed in the first three chapters, it is natural to ask the following questions:

- Can one associate with these more general filters bases of functions that play a role similar to the role played by wavelets in the orthogonal case?

- Is it still possible to estimate the quadratic norms of the signals in spite of the loss of orthogonality?

This last point is essential: The operator that maps the signal into its multiscale representation, as well as its inverse, must be continuous to ensure that the decomposition and reconstruction algorithms are stable.

We will answer these two questions by constructing explicitly unconditional bases of biorthogonal wavelets.

We have intentionally omitted from this chapter all of the multivariate constructions. It is clear that these constructions are possible, either in a trivial way by using tensor products of the univariate objects, or in a general way by considering 2^N filters and a decimation of order 2^N in the case of the analysis of $l^2(\mathbb{Z}^N)$. This is related to the multivariate constructions given in Chapter 1 and in Appendix B. Tensor product biorthogonal wavelets were used successfully by Michel Barlaud and his team for image compression algorithms (see Antonini, Barlaud, Daubechies, and Mathieu (1992)).

Our main objective in this chapter is to construct unconditional biorthogonal bases and to examine their various properties.

4.3 Unconditional biorthogonal wavelet bases

By analogy with previous constructions, we write $m_0(\omega) = \overline{H_1(\omega)}$ and $\tilde{m}_0(\omega) = H_2(\omega + \pi)$. We thus have two low-pass filters with finite impulse response, and, indeed, one should keep in mind that m_0 and $\tilde{m}_0$ are trigonometric polynomials. Relation (4.9) becomes

$$\overline{m_0(\omega)}\,\tilde{m}_0(\omega) + \overline{m_0(\omega + \pi)}\,\tilde{m}_0(\omega + \pi) = 1. \qquad (4.14)$$

The orthogonal case corresponds to the equality $m_0(\omega) = \tilde{m}_0(\omega)$.

We can now define, in the sense of tempered distributions, the two approximating functions φ and $\tilde{\varphi}$ by

$$\hat{\varphi}(\omega) = \prod_{k=1}^{+\infty} m_0(2^{-k}\omega), \qquad (4.15)$$

and

$$\hat{\tilde{\varphi}}(\omega) = \prod_{k=1}^{+\infty} \tilde{m}_0(2^{-k}\omega). \qquad (4.16)$$

We next construct the two functions ψ and $\tilde{\psi}$ that will eventually generate our multiscale biorthogonal bases. These functions are obtained from φ and $\tilde{\varphi}$ respectively. To ensure a reconstruction identity similar to the one we have in the orthogonal case (see Chapter 1), it is natural to define

$$\hat{\psi}(2\omega) = m_1(\omega)\hat{\varphi}(\omega) \quad \text{with} \quad m_1(\omega) = e^{-i\omega}\overline{\tilde{m}_0(\omega + \pi)}, \quad (4.17)$$

$$\hat{\tilde{\psi}}(2\omega) = \tilde{m}_1(\omega)\hat{\tilde{\varphi}}(\omega) \quad \text{with} \quad \tilde{m}_1(\omega) = e^{-i\omega}\overline{m_0(\omega + \pi)}. \quad (4.18)$$

These definitions yield the relations

$$\hat{\varphi}(\omega) = \overline{\tilde{m}_0(\omega)}\,\hat{\varphi}(2\omega) + \overline{\tilde{m}_1(\omega)}\,\hat{\psi}(2\omega), \qquad (4.19)$$

$$\hat{\tilde{\varphi}}(\omega) = \overline{m_0(\omega)}\,\hat{\tilde{\varphi}}(2\omega) + \overline{m_1(\omega)}\,\hat{\tilde{\psi}}(2\omega). \qquad (4.20)$$

We wish to construct unconditional bases for $L^2(\mathbb{R})$ of the type $\{\psi_{j,k}\}_{j,k\in\mathbb{Z}} = \{2^{j/2}\psi(2^j x - k)\}_{j,k\in\mathbb{Z}}$ and $\{\tilde{\psi}_{j,k}\}_{j,k\in\mathbb{Z}}$ that satisfy properties (4.3) and (4.4).

The first requirement in this program is to show that the functions φ, $\tilde{\varphi}$, ψ and $\tilde{\psi}$ are in $L^2(\mathbb{R})$. From the properties of m_0 and

$\tilde{m}_0$ we know that the products in (4.15) and (4.16) converge to continuous functions on compact sets, but we do not know the behavior of these functions at infinity. Fortunately, we can deduce from results in Chapter 3 criteria that ensure rapid enough decay at infinity to put these functions in $L^2(\mathbb{R})$.

Proposition 4.1 *Assume that the polynomials m_0 and $\tilde{m}_0$ can be factored as*

$$m_0(\omega) = \left(\frac{1+e^{i\omega}}{2}\right)^N f(\omega), \qquad (4.21)$$

and

$$\tilde{m}_0(\omega) = \left(\frac{1+e^{i\omega}}{2}\right)^{\tilde{N}} \tilde{f}(\omega), \qquad (4.22)$$

and consider their critical exponents

$$b = \inf_{j\in\mathbb{Z}}\left[(j\log 2)^{-1}\log\left(\sup_{\omega\in\mathbb{R}}\Big|\prod_{k=1}^{j}f(2^{-k}\omega)\Big|\right)\right], \qquad (4.23)$$

and $\tilde{b}$, which is similarly defined.

Then, if $N - b > \frac{1}{2}$ and $\tilde{N} - \tilde{b} > \frac{1}{2}$, the functions φ and $\tilde{\varphi}$ are in $L^2(\mathbb{R})$. More precisely, there exists an $\varepsilon > 0$ and a constant C, which depends on ε, such that

$$|\hat{\varphi}(\omega)| + |\hat{\tilde{\varphi}}(\omega)| \leq C(1+|\omega|)^{-\frac{1}{2}-\varepsilon}. \qquad (4.24)$$

The same is true for the functions ψ and $\tilde{\psi}$.

Proof. The result for φ and $\tilde{\varphi}$ follows directly from Proposition 3.3 in Chapter 3. The result for ψ and $\tilde{\psi}$ then follows from (4.17) and (4.18). We note that this criterion does not provide us with a strictly necessary condition for these functions to be in $L^2(\mathbb{R})$. □

In the rest of this section we will assume that the polynomials m_0 and $\tilde{m}_0$ satisfy the hypotheses of Proposition 4.1.

With these assumptions, we can announce the main result that will be proved in this section.

Theorem 4.1 *The families $\{\psi_{j,k}\}_{j,k\in\mathbb{Z}}$ and $\{\tilde{\psi}_{j,k}\}_{j,k\in\mathbb{Z}}$ are unconditional biorthogonal bases for $L^2(\mathbb{R})$.*

The proof of this theorem is long, but it breaks naturally into several distinct steps. A certain number of intermediate results will smooth the way toward the desired construction. The proofs of these results will use several versions of the Poisson summation formula. To avoid repeating arguments, we collect in the following lemma the results on Poisson summation that we will be using.

Lemma 4.1 *Assume that f and g are in $L^2(\mathbb{R})$ and write*

$$S(\hat{f},\hat{g})(\omega) = \sum_{l\in\mathbb{Z}} \hat{f}(\omega+2l\pi)\overline{\hat{g}(\omega+2l\pi)},$$

where the equality means that the value of $S(\hat{f},\hat{g})(\omega)$ is equal to the sum on the right when it exists. The following results give sufficient conditions for this sum to exist almost everywhere and to define an element in one of the function spaces $L^p[0,2\pi]$, $1 \leq p \leq \infty$:

i) $S(\hat{f},\hat{g})$ *is defined almost everywhere,* $S(\hat{f},\hat{g}) \in L^1[0,2\pi]$, *and*

$$\int f(x)\overline{g(x-k)}\,dx = \frac{1}{2\pi}\int_0^{2\pi} S(\hat{f},\hat{g})(\omega)e^{ik\omega}\,d\omega. \qquad (4.25)$$

ii) If g has compact support, $S(\hat{f},\hat{g}) \in L^2[0,2\pi]$, and

$$S(\hat{f},\hat{g})(\omega) = \sum_{l\in\mathbb{Z}} s_k(\hat{f},\hat{g})\,e^{-ik\omega}, \qquad (4.26)$$

where

$$s_k(\hat{f},\hat{g}) = \frac{1}{2\pi}\int \hat{f}(\omega)\overline{\hat{g}(\omega)}e^{ik\omega}\,d\omega = \int f(x)\overline{g(x-k)}\,dx$$

and the series converges to $S(\hat{f},\hat{g})$ in the sense of $L^2(\mathbb{R})$.

iii) If f and g have compact support, $S(\hat{f},\hat{g}) \in L^\infty[0,2\pi]$, and the sum in (4.26) is a trigonometric polynomial.

iv) If $|\hat{f}(\omega)| + |\hat{g}(\omega)| \leq C(1+|\omega|)^{-\frac{1}{2}-\varepsilon}$, $S(\hat{f},\hat{g}) \in L^\infty[0,2\pi]$.

v) If $|\hat{f}(\omega)| + |\hat{g}(\omega)| \leq C(1+|\omega|)^{-\frac{1}{2}-\varepsilon}$ and f and g have compact support, then $S(\hat{f},\hat{g})$ is a trigonometric polynomial.

vi) If $|\hat{g}(\omega)| \leq C(1+|\omega|)^{-\frac{1}{2}-\varepsilon}$, the conclusions of ii) hold.

Proof. 1) By integrating both sides of the inequality

$$|S(\hat{f},\hat{g})(\omega)| \leq \sum_{l\in\mathbb{Z}} |\hat{f}(\omega+2l\pi)|\,|\hat{g}(\omega+2l\pi)|$$

we see that

$$\int_0^{2\pi} |S(\hat{f}, \hat{g})(\omega)|\, dx \leq \int_0^{2\pi} \sum_{l\in\mathbb{Z}} |\hat{f}(\omega + 2l\pi)|\, |\hat{g}(\omega + 2l\pi)|\, d\omega$$

$$= \sum_{l\in\mathbb{Z}} \int_0^{2\pi} |\hat{f}(\omega + 2l\pi)|\, |\hat{g}(\omega + 2l\pi)|\, d\omega$$

$$= \int |\hat{f}(\omega)|\, |\hat{g}(\omega)|\, d\omega < +\infty$$

since $\hat{f}\hat{g} \in L^1(\mathbb{R})$. Then we can write

$$\int f(x)\overline{g(x-k)}\, dx = \frac{1}{2\pi} \int \hat{f}(\omega)\overline{\hat{g}(\omega)}\, e^{ik\omega}\, d\omega$$

$$= \frac{1}{2\pi} \sum_{l\in\mathbb{Z}} \int_0^{2\pi} \hat{f}(\omega + 2l\pi)\, \overline{\hat{g}(\omega + 2l\pi)}\, e^{ik\omega}\, d\omega$$

$$= \frac{1}{2\pi} \int_0^{2\pi} S(\hat{f}, \hat{g})(\omega)\, e^{ik\omega}\, d\omega,$$

where everything in sight converges absolutely. We prove iii) next.

iii) Since f and g have compact support, the integral on the left side of (4.25) is zero except for a finite number of integers k. Thus, $S(\hat{f}, \hat{g})$, as an element of $L^1[0, 2\pi]$, is equal to a trigonometric polynomial, and this means that $S(\hat{f}, \hat{g})$ is in $L^\infty[0, 2\pi]$.

ii) In this case we use the Schwarz inequality and write

$$|S(\hat{f}, \hat{g})(\omega)|^2 \leq \sum_{l\in\mathbb{Z}} |\hat{f}(\omega + 2l\pi)|^2 \sum_{l\in\mathbb{Z}} |\hat{g}(\omega + 2l\pi)|^2$$

$$= S(\hat{f}, \hat{f})(\omega)\, S(\hat{g}, \hat{g})(\omega).$$

$S(\hat{f}, \hat{f})$ is in $L^1[0, 2\pi]$ by i), and $S(\hat{g}, \hat{g})$ is in $L^\infty[0, 2\pi]$ by iii). Thus, the left-hand side is in $L^1[0, 2\pi]$ and $S(\hat{f}, \hat{g})$ is in $L^2[0, 2\pi]$. The other part of ii) follows from the computations we have already made.

iv) By the hypothesis,

$$|\hat{f}(\omega + 2l\pi)|\, |\hat{g}(\omega + 2l\pi)| \leq C'(1 + |\omega + 2l\pi|)^{-1-2\varepsilon}.$$

If $\omega \in [0, 2\pi]$, the right-hand side is dominated by $C'(1+|2l\pi|)^{-1-2\varepsilon}$ for $l \geq 0$ and by $C'(1 + |2\pi + 2l\pi|)^{-1-2\varepsilon}$ for $l < 0$. Thus the series that defines $S(\hat{f}, \hat{g})$ converges uniformly on $[0, 2\pi]$ to a bounded function, which means that $S(\hat{f}, \hat{g})$ is in $L^\infty[0, 2\pi]$.

v) Since f and g have compact support, $\hat{f}$ and $\hat{g}$ are continuous (in fact, C^∞). The terms of the series that defines $S(\hat{f}, \hat{g})$ are continuous functions, and the uniform convergence implies that the limit is also continuous. It is clear that (4.26) is true for all $\omega \in [0, 2\pi]$ so that $S(\hat{f}, \hat{g})$ is indeed a trigonometric polynomial.

vi) The proof is the same as that of ii) except here $S(\hat{g}, \hat{g})$ is in $L^\infty[0, 2\pi]$ by iv). □

As a first step in the proof of the theorem itself, we introduce formally the approximations and details at scale 2^j of an element f in $L^2(\mathbb{R})$. We write

$$\begin{aligned} T_j f &= \sum_{k \in \mathbb{Z}} 2^j \langle f, \tilde{\varphi}(2^j x - k) \rangle \varphi(2^j x - k) \\ &= \sum_{k \in \mathbb{Z}} \langle f, \tilde{\varphi}_{j,k} \rangle \varphi_{j,k}, \end{aligned} \qquad (4.27)$$

and

$$\begin{aligned} \Delta_j f &= \sum_{k \in \mathbb{Z}} 2^j \langle f, \tilde{\psi}(2^j x - k) \rangle \psi(2^j x - k) \\ &= \sum_{k \in \mathbb{Z}} \langle f, \tilde{\psi}_{j,k} \rangle \psi_{j,k}. \end{aligned} \qquad (4.28)$$

$\tilde{T}_j$ and $\tilde{\Delta}_j$ are defined similarly by interchanging φ and $\tilde{\varphi}$ as well as ψ and $\tilde{\psi}$. The first result tells us more about these new tools.

Proposition 4.2 $T_j, \Delta_j, \tilde{T}_j,$ and $\tilde{\Delta}_j$ are bounded operators on $L^2(\mathbb{R})$. Furthermore, their norms are independent of j.

Proof. We will prove the result for T_j. The proof is exactly the same for the three other operators. Note first that

$$T_j f(y)(x) = T_0 f(2^{-j} y)(2^j x). \qquad (4.29)$$

It is thus sufficient to show that T_0 is a bounded operator, for from this it follows by changing variables that

$$\|T_j\| = \|T_0\| \qquad \text{for all } j \text{ in } \mathbb{Z}. \qquad (4.30)$$

The continuity of T_0 comes essentially from the fact that the functions φ and $\tilde{\varphi}$ are in $L^2(\mathbb{R})$ and have compact support. (Recall the argument in Section 2.4 about the compact support of these functions.)

Indeed, by Lemma 4.1, v), the two functions α and $\tilde{\alpha}$ defined by $\alpha(\omega) = \sum_{l \in \mathbb{Z}} |\hat{\varphi}(\omega + 2l\pi)|^2$ and $\tilde{\alpha}(\omega) = \sum_{l \in \mathbb{Z}} |\hat{\tilde{\varphi}}(\omega + 2l\pi)|^2$ are trigonometric polynomials whose Fourier coefficients $\{c_k\}$ and $\{\tilde{c}_k\}$

are given by $c_k = \langle \varphi(x), \varphi(x-k) \rangle$ and $\tilde{c}_k = \langle \tilde{\varphi}(x), \tilde{\varphi}(x-k) \rangle$. Thus, α and $\tilde{\alpha}$ are bounded.

The proof proceeds by estimating the L^2 norm of

$$T_0 f = \sum_{k \in \mathbb{Z}} \langle f, \tilde{\varphi}(x-k) \rangle \varphi(x-k)$$

in two steps.

- For any sequence $\{e_k\}$ in $l^2(\mathbb{Z})$, the function

$$g(x) = \sum_{k \in \mathbb{Z}} e_k \varphi(x-k)$$

is in $L^2(\mathbb{R})$, and $\hat{g}(\omega) = E(\omega)\hat{\varphi}(\omega)$, where $E(\omega) = \sum e_k e^{-ik\omega}$. Hence,

$$\begin{aligned}
\int |\sum_{k \in \mathbb{Z}} e_k \varphi(x-k)|^2 \, dx &= \frac{1}{2\pi} \int |E(\omega)|^2 |\hat{\varphi}(\omega)|^2 \, d\omega \\
&= \frac{1}{2\pi} \sum_{l \in \mathbb{Z}} \int_{-\pi}^{\pi} |E(\omega)|^2 |\hat{\varphi}(\omega + 2l\pi)|^2 \, d\omega \\
&= \frac{1}{2\pi} \int_{-\pi}^{\pi} |E(\omega)|^2 \alpha(\omega) \, d\omega \\
&\leq \sup_{\omega \in \mathbb{R}} \alpha(\omega) \sum_{k \in \mathbb{Z}} |e_k|^2.
\end{aligned}$$

- We show that the sequence $\langle f, \tilde{\varphi}(x-k) \rangle$ belongs to $l^2(\mathbb{Z})$ and estimate its norm with the following computation.

$$\begin{aligned}
\sum_{k \in \mathbb{Z}} |\langle f, \tilde{\varphi}(x-k) \rangle|^2 &= \frac{1}{4\pi^2} \sum_{k \in \mathbb{Z}} \left| \int \hat{f}(\omega) \overline{\hat{\tilde{\varphi}}(\omega)} e^{ik\omega} \, d\omega \right|^2 \\
&= \frac{1}{2\pi} \int_{-\pi}^{\pi} \left| \sum_{k \in \mathbb{Z}} \hat{f}(\omega + 2l\pi) \overline{\hat{\tilde{\varphi}}(\omega + 2l\pi)} \right|^2 d\omega \\
&\leq \frac{1}{2\pi} \int_{-\pi}^{\pi} \left(\sum_{k \in \mathbb{Z}} |\hat{f}(\omega + 2l\pi)|^2 \right) \left(\sum_{l \in \mathbb{Z}} |\hat{\tilde{\varphi}}(\omega + 2l\pi)|^2 \right) d\omega \\
&\leq \sup_{\omega \in \mathbb{R}} \tilde{\alpha}(\omega) \int |f(x)|^2 \, dx.
\end{aligned}$$
(4.31)

We have used Lemma 4.1 (either ii) or iv)) to obtain the second equality.

These two estimates show that
$$\int |T_0 f(x)|^2 \, dx \leq \sup_{\omega \in \mathbb{R}} \alpha(\omega) \sup_{\omega \in \mathbb{R}} \tilde{\alpha}(\omega) \int |f(x)|^2 \, dx.$$
This, combined with (4.30), proves that
$$\|T_j\| \leq [\sup \alpha \ \sup \tilde{\alpha}]^{1/2} \qquad (4.32)$$
for all j in $\mathbb{Z}$. □

We remark that $\tilde{T}_j$ is the dual of T_j, and similarly, that $\tilde{\Delta}_j$ is the dual of Δ_j. Ultimately, we will see that the operators T_j, $\tilde{T}_j$, Δ_j, and $\tilde{\Delta}_j$ are non-orthogonal projections on the approximation and detail subspaces.

It is time to use the inter-scale transition relations, which are related to the properties of the functions m_0 and $\tilde{m}_0$, to show the connections among the four operators $T_j, \tilde{T}_j, \Delta_j,$ and $\tilde{\Delta}_j$. Based on formulas (4.14) to (4.20), we will prove a result that is exactly the same as the corresponding result for the orthogonal case. This states that the approximation at a given scale is the sum of the approximation and the details at the next larger scale, which is the given scale multiplied by 2.

Proposition 4.3 *The operators $T_j, \tilde{T}_j, \Delta_j,$ and $\tilde{\Delta}_j$ satisfy the following relations for all $j \in \mathbb{Z}$:*
$$T_j = T_{j-1} + \Delta_{j-1}, \qquad (4.33)$$
and
$$\tilde{T}_j = \tilde{T}_{j-1} + \tilde{\Delta}_{j-1}. \qquad (4.34)$$

Proof. Since (4.33) and (4.34) are equivalent by duality, it is sufficient to prove (4.33). From the scaling relation (4.29), we see that it is also sufficient to prove the equality $T_0 = T_{-1} + \Delta_{-1}$, which by (4.27) and (4.28) is

$$\sum_{k \in \mathbb{Z}} \langle f, \tilde{\varphi}(x-k) \rangle \, \varphi(x-k)$$
$$= \frac{1}{2} \sum_{k \in \mathbb{Z}} \left[\langle f, \tilde{\varphi}(x/2 - k) \rangle \varphi(x/2 \quad k) \right.$$
$$\left. + \langle f, \tilde{\psi}(x/2 - k) \rangle \psi(x/2 - k) \right].$$

Furthermore, since we have shown that these operators are continuous on $L^2(\mathbb{R})$, it is sufficient to show that this last relation holds on a dense subset of $L^2(\mathbb{R})$. Thus, we may assume that f has compact support and that $|\hat{f}(\omega)| \leq C(1+|\omega|)^{-1}$. This assumption about f guarantees, by Lemma 4.1, that all the equations in the proof — (4.35) through (4.40) — hold for all ω. (Alternatively, one must interpret these expressions in the sense of L^2.)

Define the sequences $s_k = \langle f, \tilde{\varphi}(x-k)\rangle$, $s_{1,k} = \langle f, \tilde{\varphi}(x/2-k)\rangle$, and $d_{1,k} = \langle f, \tilde{\psi}(x/2-k)\rangle$ and denote their discrete Fourier transforms by S, S_1, and D_1. Then the last identity can be expressed as

$$S(\omega)\hat{\varphi}(\omega) = S_1(2\omega)\hat{\varphi}(2\omega) + D_1(2\omega)\hat{\psi}(2\omega). \tag{4.35}$$

By (4.15) and (4.17), this will be satisfied if

$$S(\omega) = m_0(\omega)S_1(2\omega) + m_1(\omega)D_1(2\omega). \tag{4.36}$$

But this last equality is nothing more than the formula for reconstructing the signal $S(\omega)$ from $S_1(2\omega)$ and $D_1(2\omega)$ using the algorithm described in Section 4.2, except that here we have added the factor $e^{-i\omega}$ in the definition of m_1. This means that the decimation in the calculation of D_1 is similar to that of S_1: the odd coefficients of both $S_1(2\omega)$ and $D_1(2\omega)$ are zero.

To verify (4.36), it will be sufficient to express S_1 and D_1 in terms of S, and this must correspond to the subband decomposition previously described.

We begin by developing an expression for $S(\omega)$:

$$\begin{aligned} S(\omega) &= \sum_{k\in\mathbb{Z}} \langle f(x), \tilde{\varphi}(x-k)\rangle e^{-ik\omega} \\ &= \frac{1}{2\pi} \sum_{k\in\mathbb{Z}} \langle \hat{f}(\xi), \widehat{\tilde{\varphi}}(\xi)e^{-ik\xi}\rangle e^{-ik\omega}. \end{aligned}$$

We use Lemma 4.1 to get

$$S(\omega) = \sum_{l\in\mathbb{Z}} \hat{f}(\omega+2l\pi)\,\overline{\widehat{\tilde{\varphi}}(\omega+2l\pi)}. \tag{4.37}$$

Similarly, we have

$$S_1(2\omega) = \frac{1}{\pi}\sum_{k\in\mathbb{Z}} \langle \hat{f}(\xi), \widehat{\tilde{\varphi}}(2\xi)e^{-i2k\xi}\rangle e^{-i2k\omega}.$$

We use Lemma 4.1 again, but this time we have $\hat{g}(\omega) = \widehat{\tilde{\varphi}}(2\omega)$. $S(\hat{f},\hat{g})$ is then π-periodic and $S(\hat{f},\hat{g})(\omega) = \sum_{k\in\mathbb{Z}} s_{2k}(\hat{f},\hat{g})\,e^{-i2k\omega}$, where $s_{2k}(\hat{f},\hat{g})$ is equal to $\frac{1}{\pi}\int \hat{f}(\omega)\overline{\hat{\varphi}(2\omega)}e^{i2k\omega}\,d\omega$. Thus,

$$S_1(2\omega) = \sum_{l\in\mathbb{Z}} \hat{f}(\omega+l\pi)\,\overline{\widehat{\tilde{\varphi}}(2(\omega+l\pi))}. \tag{4.38}$$

From (4.16) and (4.37) we see that

$$\overline{\tilde{m}_0(\omega)}\,S(\omega) = \sum_{l\text{ even}} \hat{f}(\omega+l\pi)\,\overline{\widehat{\tilde{\varphi}}(2(\omega+l\pi))},$$

and

$$\overline{\tilde{m}_0(\omega+\pi)}\,S(\omega+\pi) = \sum_{l\text{ odd}} \hat{f}(\omega+l\pi)\,\overline{\widehat{\tilde{\varphi}}(2(\omega+l\pi))}.$$

These relations and (4.38) show that

$$S_1(2\omega) = \overline{\tilde{m}_0(\omega)}\,S(\omega) + \overline{\tilde{m}_0(\omega+\pi)}\,S(\omega+\pi). \tag{4.39}$$

A similar computation shows that

$$D_1(2\omega) = \overline{\tilde{m}_1(\omega)}\,S(\omega) + \overline{\tilde{m}_1(\omega+\pi)}\,S(\omega+\pi). \tag{4.40}$$

These are effectively the decomposition formulas with decimation on the odd coefficients in both cases.

It remains to verify (4.36). From (4.39) and (4.40) we get

$$S_1(2\omega)m_0(\omega) + D_1(2\omega)m_1(\omega)$$
$$= S(\omega)\left[m_0(\omega)\overline{\tilde{m}_0(\omega)} + m_1(\omega)\overline{\tilde{m}_1(\omega)}\right]$$
$$+ S(\omega+\pi)\left[m_0(\omega)\overline{\tilde{m}_0(\omega+\pi)} + m_1(\omega)\overline{\tilde{m}_1(\omega+\pi)}\right],$$

and by hypothesis (4.14) the right-hand side is exactly the original signal $S(\omega)$. The proves the proposition. $\square$

An immediate consequence of this result is obtained by iterating it over several scales. Thus for $J > 0$ we have

$$T_J = T_{-J} + \sum_{j=-J}^{J-1} \Delta_j. \tag{4.41}$$

The next step examines the behavior of T_J as $|J|$ tends to $+\infty$.

Proposition 4.4 *For all f in $L^2(\mathbb{R})$,*

$$\lim_{j\to+\infty} T_j f = f,\ i.e.,\ \lim_{j\to+\infty}\int |T_j f(x) - f(x)|^2\,dx = 0, \tag{4.42}$$

$$\lim_{j\to-\infty} T_j f = 0,\ i.e.,\ \lim_{j\to-\infty}\int |T_j f(x)|^2\,dx = 0, \tag{4.43}$$

and consequently

$$f = \lim_{N\to+\infty} \sum_{j=-N}^{N} \sum_{k\in\mathbb{Z}} \langle f, \tilde{\psi}_{j,k}\rangle \psi_{j,k}. \qquad (4.44)$$

Proof. We use a remark that we made in Chapter 1: Each function f in $L^2(\mathbb{R})$ can be approximated arbitrarily closely by a linear combination of characteristic functions of bounded intervals.

Let g be one of these step functions satisfying $\|f - g\| \leq \varepsilon$. Then by Proposition 4.2 we have

$$\|T_j f\| = \|T_j(f - g + g)\| \leq \|T_j g\| + \|T_0\|\varepsilon, \qquad (4.45)$$

and

$$\|T_j f - f\| \leq \|T_j g - g\| + (\|T_0\| + 1)\varepsilon. \qquad (4.46)$$

It is thus sufficient to prove (4.42) and (4.43) for $f = \mathbf{1}_{[a,b]}$.

a) We first prove (4.43). From the proof of Proposition 4.2 we know that

$$\|T_j f\|^2 \leq \sup_{\omega\in\mathbb{R}} \alpha(\omega) \sum_{k\in\mathbb{Z}} |\langle f, \tilde{\varphi}_{j,k}\rangle|^2.$$

We now refer to the proof of part (b) of Theorem 1.1 where we showed that the sum $\sum_{k\in\mathbb{Z}} |\langle \mathbf{1}_{[a,b]}, \varphi_{j,k}\rangle|^2$ tends to zero as j tends to $-\infty$. That argument depended only on the fact that φ was in $L^2(\mathbb{R})$. Consequently, the same argument proves that $\sum_{k\in\mathbb{Z}} |\langle f, \tilde{\varphi}_{j,k}\rangle|^2$ tends to zero as j tends to $-\infty$.

b) To prove (4.42), we are going to evaluate $T_j f$ directly, always assuming that $f = \mathbf{1}_{[a,b]}$. It will then become quite clear that $T_j f$ tends to f in the sense of $L^2(\mathbb{R})$ as j tends to $+\infty$.

However, before proceeding with the proof itself, it will be useful to review some of the properties of φ and $\tilde{\varphi}$.

• From Proposition 4.1, we know that φ and $\tilde{\varphi}$ are in $L^2(\mathbb{R})$. We noted in Section 2.4 that φ, as a distribution, has compact support. In fact, if $m_0(\omega) = \sum_{n=N_1}^{N_2} h_n e^{-in\omega}$, then

$$\varphi = \binom{+\infty}{*}_{k=1} \left(\sum_{n=N_1}^{N_2} h_n \delta(2^{-k}n)\right) = \binom{+\infty}{*}_{k=1} u_k. \qquad (4.47)$$

This convolution product converges in the sense of distributions to φ. Recalling that $\delta(a) * \delta(b) = \delta(a+b)$, we see that the partial products can be represented as

$$\left(*\right)_{j=1}^{J} u_j = \left(*\right)_{j=1}^{J} \sum_{n_j=N_1}^{N_2} h_{n_j}\, \delta(2^{-j} n_j)$$

$$= \sum_{n_1=N_1}^{N_2} \sum_{n_2=N_1}^{N_2} \cdots \sum_{n_J=N_1}^{N_2} h_{n_1} h_{n_2} \cdots h_{n_J} \delta\Big(\frac{n_1}{2} + \frac{n_2}{4} + \cdots + \frac{n_J}{2^J}\Big).$$

We immediately read two facts from this last equation: The distribution $(*)_{j=1}^{J} u_j$ consists of δ-functions at the points $2^{-J} n$, where $n = N_1(2^J - 1), \ldots, N_2(2^J - 1)$, and, consequently, in the limit, the support of the distribution is in $[N_1, N_2]$. Also, the total mass of these δ-functions is

$$\sum_{n_1=N_1}^{N_2} \sum_{n_2=N_1}^{N_2} \cdots \sum_{n_J=N_1}^{N_2} h_{n_1} h_{n_2} \cdots h_{n_J} = \Big(\sum_{n=N_1}^{N_2} h_n\Big)^J = 1 \quad (4.48)$$

because $m_0(0) = 1$. This means that the support of φ is in $[N_1, N_2]$ (and hence, in our case, that φ is in $L^1(\mathbb{R})$) and that its integral is 1. This is true for $\tilde{\varphi}$ by the same argument, so we have

$$\int \varphi(x)\, dx = \int \tilde{\varphi}(x)\, dx = 1. \quad (4.49)$$

- Our assumptions about m_0 and $\tilde{m}_0$ imply that the expressions (4.15) and (4.16) for φ and $\tilde{\varphi}$ converge uniformly on compact sets and are thus well defined for all ω. Also, $m_0(0) = \tilde{m}_0(0) = 1$ and $m_0(\pi) = \tilde{m}_0(\pi) = 0$. From this we deduce that

$$\hat{\varphi}(2k\pi) = \hat{\tilde{\varphi}}(2k\pi) = \begin{cases} 1 & \text{if } k = 0 \\ 0 & \text{if } k \neq 0. \end{cases} \quad (4.50)$$

Indeed, if k is not zero, then the products (4.15) and (4.16) always contain a null factor coming from an odd multiple of π.

- Since φ and $\tilde{\varphi}$ have compact support, we will assume that they are zero outside some interval $[-s, s]$. Let $g(x) = \sum_{k \in \mathbb{Z}} \varphi(x-k)$ and $\tilde{g}(x) = \sum_{k \in \mathbb{Z}} \tilde{\varphi}(x - k)$. For each fixed x, or for all $x \in [0, 1]$, there are only a finite number of non-zero terms in these sums, namely the k for which $x - k \in [-s, s]$. Hence, g and $\tilde{g}$ are defined almost everywhere on $\mathbb{R}$, they have period 1, and they are in $L^1[0, 1]$ and

$L^2[0,1]$. By taking the Fourier transform of g we see that

$$\int_0^1 g(x)\, e^{-i2n\pi x}\, dx = \int \varphi(x)\, e^{-i2n\pi x}\, dx = \hat{\varphi}(2n\pi),$$

and similarly for $\tilde{g}$. Consequently, by (4.50),

$$\sum_{k \in \mathbb{Z}} \varphi(x-k) = \sum_{k \in \mathbb{Z}} \tilde{\varphi}(x-k) = 1 \qquad (4.51)$$

almost everywhere. (This also establishes (4.49) independently of the 'mass equals 1' argument given above.) Note that, at this point, φ and $\tilde{\varphi}$ exist only as elements of $L^2(\mathbb{R})$ so, strictly speaking, it is necessary to continue the 'almost everywhere' statement. We will not do this, but it should be understood in what follows.

We now return to the proof. Note that for a fixed x, the sum in the expression $T_j f(x) = 2^{j/2} \sum_{k \in \mathbb{Z}} \langle f, \tilde{\varphi}_{j,k} \rangle \varphi(2^j x - k)$ involves only those k for which $2^j x - k \in [-s, s]$. We begin by analyzing the scalar products

$$\langle f, \tilde{\varphi}_{j,k} \rangle = \int_a^b 2^{j/2}\, \overline{\tilde{\varphi}(2^j x - k)}\, dx = 2^{-j/2} \int_{2^j a - k}^{2^j b - k} \overline{\tilde{\varphi}(x)}\, dx$$

that appear in $T_j f$. For j sufficiently large, $2^{-j+2} s < b - a$, and we have three cases for the evaluation of these scalar products.

Case 1: The number $2^{-j} k$ is in the interval $[a + 2^{-j} s, b - 2^{-j} s]$. Then $[-s, s]$ is contained in the interval $[2^j a - k, 2^j b - k]$, and we see from (4.49) that $\langle f, \tilde{\varphi}_{j,k} \rangle = 2^{-j/2}$.

Case 2: $2^{-j} k$ is outside $[a - 2^{-j} s, b + 2^{-j} s]$, in which case $[-s, s]$ and $[2^j a - k, 2^j b - k]$ are disjoint and $\langle f, \tilde{\varphi}_{j,k} \rangle = 0$.

Case 3: $2^{-j} k$ is in one of the intervals of length $2^{-j+1} s$ centered at a and b.

From all of this and the preceding remarks we deduce the following:

- If x is in the interval $[a + 2^{-j+1} s, b - 2^{-j+1} s]$, the condition $2^j x - k \in [-s, s]$ implies that we are in Case 1, and hence the sum for $T_j f(x)$ involves only terms for which $\langle f, \tilde{\varphi}_{j,k} \rangle = 2^{-j/2}$. It can thus be written as

$$T_j f(x) = \sum_{k \in \mathbb{Z}} 2^{-j/2}\, \varphi_{j,k}(x) = \sum_{k \in \mathbb{Z}} \varphi(2^j x - k) = 1. \qquad (4.52)$$

- If x is outside the interval $[a + 2^{-j+1} s, b - 2^{-j+1} s]$, the condition

$2^j x - k \in [-s,s]$ implies that we are in Case 2. Hence the sum for $T_j f(x)$ involves only null terms, and we have $T_j f(x) = 0$.

Thus, on these two intervals, $T_j f(x)$ equals $f(x)$ almost everywhere, and we have

$$\int |T_j f(x) - f(x)|^2\, dx = \int_{R_j} |T_j f(x) - f(x)|^2\, dx,$$

where R_j is the residual part around a and b. Hence, the last step is to evaluate the integral $\int_{R_j} |T_j f(x) - f(x)|^2\, dx$.

Consider $R_{j-1} = [a-2^{-j+2}s, a+2^{-j+2}s] \cup [b-2^{-j+2}s, b+2^{-j+2}s]$. Clearly, $R_j \subset R_{j-1}$. Define $f_j(x) = f(x)\mathbf{1}_{R_{j-1}}(x)$. Then for x in R_j, we have the following situation:

- $T_j f(x) = T_j f_j(x)$.

This can be seen by writing

$$T_j f(x) - T_j f_j(x) = \sum_{k \in \mathbb{Z}} 2^{-j/2} \left[\int_{2^j a + 4s - k}^{2^j b - 4s - k} \overline{\tilde{\varphi}(x)}\, dx \right] \varphi(2^j x - k)$$

and noting that, if $x \in R_j$ and $2^j x - k \in [-s,s]$, then the range of integration and the support of $\tilde{\varphi}$ are disjoint.

- $f(x) = f_j(x)$ since $R_j \subset R_{j-1}$.

Thus we have

$$\begin{aligned}\int_{R_j} |T_j f(x) - f(x)|^2\, dx &= \int_{R_j} |T_j f_j(x) - f_j(x)|^2\, dx \\ &\leq \int_{\mathbb{R}} |T_j f_j(x) - f_j(x)|^2\, dx \\ &\leq (\|T_0\| + 1)^2 \|f_j\|^2.\end{aligned}$$

Finally, it is clear that the L^2 norm of f_j tends to 0 when j tends to $+\infty$, and this completes the proof of Proposition 4.4. □

This proof reveals a phenomenon that is common to all wavelets with compact support, whether they are orthonormal or biorthogonal: The approximation at the scale 2^j of a portion of a function where it is constant is exact, except for two intervals with length proportional to 2^{-j}. This also shows that, in this approximation setting, the 'Gibbs phenomenon' is localized in space, which is contrary to the standard Gibbs phenomenon appearing in the approximation by a band-limited function (see Oppenheim and Schafer

(1975)). This will be illustrated in the applications to image compression that are presented in Chapter 5.

By assuming higher-order cancellation relations, that is to say, $\left(\frac{d}{d\omega}\right)^k(m_0)(\pi) = 0$ for $k = 0, \ldots, N$, one can prove a similar result for the approximation of functions that behave locally like polynomials. By this we mean that polynomials of degree N can be expressed locally as linear combinations of the $\varphi(\cdot - k)$, $k \in \mathbb{Z}$. The equivalence between the reproduction of polynomials of degree N and the flatness of m_0 at π, or equivalently of $\hat{\varphi}$ at $2k\pi$, $k \in \mathbb{Z} \setminus \{0\}$, is known as the Strang–Fix condition of order N (see Fix and Strang (1969)). These conditions are particularly useful in approximation theory since smooth functions are locally close to polynomials and can thus be well approximated by the spaces V_j. (For a description of smoothness classes in the context of approximation theory, see DeVore and Lorentz (1993).)

Formula (4.44) provides a way to decompose and reconstruct (analysis and synthesis) any element f in $L^2(\mathbb{R})$. Note, however, that this in no way means that the families $\{\psi_{j,k}\}$ and $\{\tilde{\psi}_{j,k}\}$ are unconditional biorthogonal bases.

The following step consists in showing that the families $\{\psi_{j,k}\}$ and $\{\tilde{\psi}_{j,k}\}$ satisfy duality formulas like (4.1). We begin by establishing a result that will reduce this duality to that of the families $\{\varphi(x-k)\}_{k\in\mathbb{Z}}$ and $\{\tilde{\varphi}(x-k)\}_{k\in\mathbb{Z}}$.

Proposition 4.5 *Assume that*

$$\langle \varphi(x-k), \tilde{\varphi}(x-k') \rangle = \delta_{k,k'} \tag{4.53}$$

for all integers k and k'. Then, for all integers k, k', j, and j',

$$\langle \psi_{j,k}, \tilde{\psi}_{j',k'} \rangle = \delta_{j,j'} \delta_{k,k'}. \tag{4.54}$$

Proof. The relations in (4.53) are equivalent to saying that

$$\sum_{l\in\mathbb{Z}} \overline{\hat{\varphi}(\omega+2l\pi)} \, \hat{\tilde{\varphi}}(\omega+2l\pi) = 1, \tag{4.55}$$

which, from our hypotheses and Lemma 4.1, holds for all ω.

We begin by establishing the following three equalities:

$$\frac{1}{2} \langle \psi\!\left(\frac{x}{2}-k\right), \tilde{\psi}\!\left(\frac{x}{2}-k'\right) \rangle = \delta_{k,k'}, \tag{4.56}$$

$$\frac{1}{2} \langle \psi\!\left(\frac{x}{2}-k\right), \tilde{\varphi}\!\left(\frac{x}{2}-k'\right) \rangle = 0, \tag{4.57}$$

and
$$\frac{1}{2}\langle \varphi\big(\frac{x}{2}-k\big), \tilde{\psi}\big(\frac{x}{2}-k'\big)\rangle = 0. \tag{4.58}$$

For this, we examine the corresponding summation formulas

$$\beta(\omega) = \sum_{l\in\mathbb{Z}} \overline{\hat{\psi}(2\omega+2l\pi)}\,\widehat{\tilde{\psi}}(2\omega+2l\pi), \tag{4.59}$$

and

$$\gamma(\omega) = \sum_{l\in\mathbb{Z}} \overline{\hat{\psi}(2\omega+2l\pi)}\,\widehat{\tilde{\varphi}}(2\omega+2l\pi), \tag{4.60}$$

which, again by Lemma 4.1, are trigonometric polynomials.

By substituting the transition formulas (4.17) and (4.18) in (4.59) and using (4.16) and (4.17) in (4.60), we see that

$$\beta(\omega) = \overline{m_1(\omega)}\,\tilde{m}_1(\omega) + \overline{m_1(\omega+\pi)}\,\tilde{m}_1(\omega+\pi) \tag{4.61}$$

and

$$\gamma(\omega) = \overline{m_1(\omega)}\,\tilde{m}_0(\omega) + \overline{m_1(\omega+\pi)}\,\tilde{m}_0(\omega+\pi) \tag{4.62}$$

for all ω. From the hypotheses, this implies that $\beta(\omega) \equiv 1$ and $\gamma(\omega) \equiv 0$. These statements are equivalent, by taking Fourier transforms, to (4.56) and (4.57) respectively. A similar computation proves (4.58).

For the next step, we denote by V_j, W_j, $\tilde{V}_j$, and $\tilde{W}_j$ the subspaces generated respectively by the families $\{\varphi_{j,k}\}_{k\in\mathbb{Z}}$, $\{\psi_{j,k}\}_{k\in\mathbb{Z}}$, $\{\tilde{\varphi}_{j,k}\}_{k\in\mathbb{Z}}$, and $\{\tilde{\psi}_{j,k}\}_{k\in\mathbb{Z}}$. By construction, it is clear that we have the inclusions $V_j \subset V_{j+1}$, $W_j \subset V_{j+1}$, $\tilde{V}_j \subset \tilde{V}_{j+1}$, and $\tilde{W}_j \subset \tilde{V}_{j+1}$. Furthermore, X_{j+1} is derived from X_j by a dilation by a factor 2.

Since (4.53), (4.56), (4.57), and (4.58) remain true under the change of scale $x \mapsto 2^j x$ for $j \in \mathbb{Z}$, we can recapitulate our results as follows:

- The generating families of V_j and $\tilde{V}_j$ satisfy the relation

$$\langle \varphi_{j,k}, \tilde{\varphi}_{j,k'}\rangle = \delta_{k,k'}. \tag{4.63}$$

- Similarly, for the subspaces W_j and $\tilde{W}_j$ we have

$$\langle \psi_{j,k}, \tilde{\psi}_{j,k'}\rangle = \delta_{k,k'}. \tag{4.64}$$

- The subspaces V_j and $\tilde{W}_j$ are orthogonal as are W_j and $\tilde{V}_j$:

$$\langle \varphi_{j,k}, \tilde{\psi}_{j,k'}\rangle = \langle \psi_{j,k}, \tilde{\varphi}_{j,k'}\rangle = 0. \tag{4.65}$$

Thus the subspace W_j is orthogonal to all of the $\tilde{V}_l$ for $l \leq j$, and consequently, to all of the $\tilde{W}_l$ for $l < j$. From this we conclude that W_j and $\tilde{W}_{j'}$ are orthogonal when $j \neq j'$. But this can be expressed by

$$\langle \psi_{j,k}, \tilde{\psi}_{j',k'} \rangle = \delta_{j,j'} \, \delta_{k,k'}, \tag{4.66}$$

which proves the proposition. □

The operator $T_j f = \sum_{k \in \mathbb{Z}} \langle f, \tilde{\varphi}_{j,k} \rangle \varphi_{j,k}$ (respectively Δ_j, $\tilde{T}_j$, $\tilde{\Delta}_j$) is, as a consequence of this result, a non-orthogonal projection onto V_j (respectively W_j, $\tilde{V}_j$, $\tilde{W}_j$) parallel to W_j (respectively V_j, $\tilde{W}_j$, $\tilde{V}_j$), and, from Proposition 4.3 we have

$$V_j = V_{j-1} \oplus W_{j-1} \quad \text{and} \quad \tilde{V}_j = \tilde{V}_{j-1} \oplus \tilde{W}_{j-1}. \tag{4.67}$$

We now pass to the result that will guarantee that all of these duality relations hold.

Proposition 4.6 *The families $\{\varphi(x-k)\}_{k \in \mathbb{Z}}$ and $\{\tilde{\varphi}(x-k)\}_{k \in \mathbb{Z}}$ satisfy the relations*

$$\langle \varphi(x-k), \tilde{\varphi}(x-k') \rangle = \delta_{k,k'} \tag{4.68}$$

for all integers k and k'.

Proof. The proof follows closely arguments we have already made in Chapters 2 and 3. We introduce the truncated products

$$\hat{h}_n(\omega) = \prod_{k=1}^{n} m_0(2^{-k}\omega) \, \mathbf{1}_{[-2^n \pi, 2^n \pi]} \tag{4.69}$$

and

$$\widehat{\tilde{h}}_n(\omega) = \prod_{k=1}^{n} \tilde{m}_0(2^{-k}\omega) \, \mathbf{1}_{[-2^n \pi, 2^n \pi]} \tag{4.70}$$

for $n > 0$ and define $\hat{h}_0(\omega) = \widehat{\tilde{h}}_0(\omega) = \mathbf{1}_{[-\pi, \pi]}$. (Compare with $\hat{f}_n$ in the proof of Theorem 2.1.) We will show that the families $\{h_n(x-k)\}_{k \in \mathbb{Z}}$ and $\{\tilde{h}_n(x-k)\}_{k \in \mathbb{Z}}$ satisfy (4.68) by following the same reasoning that was used in the proof of Theorem 2.1 in Chapter 2.

The first step is to prove that $\int_{\mathbb{R}} \overline{\hat{h}_n(\omega)} \, \widehat{\tilde{h}}_n(\omega) \, e^{il\omega} \, d\omega = 2\pi \delta_{0,l}$ for all $n \geq 0$. This is clearly true for $n = 0$:

$$\int_{\mathbb{R}} \overline{\hat{h}_0(\omega)} \, \widehat{\tilde{h}}_0(\omega) \, e^{il\omega} \, d\omega = \int_{-\pi}^{\pi} e^{i2l\omega} \, d\omega = 2\pi \delta_{0,l}.$$

UNCONDITIONAL BIORTHOGONAL WAVELET BASES

For $n \geq 0$, we have

$$\int_{\mathbb{R}} \overline{\hat{h}_n(\omega)}\, \hat{\tilde{h}}_n(\omega)\, e^{il\omega}\, d\omega$$

$$= \int_{-2^n\pi}^{2^n\pi} \left(\prod_{k=1}^{n} \overline{m_0(2^{-k}\omega)}\, \tilde{m}_0(2^{-k}\omega) \right) e^{il\omega}\, d\omega$$

$$= 2^{n+1} \int_{-\pi/2}^{\pi/2} \left(\prod_{k=1}^{n} \overline{m_0(2^{k}\omega)}\, \tilde{m}_0(2^{k}\omega) \right) e^{i2^{n+1}l\omega}\, d\omega$$

by replacing the variable ω with $2^{n+1}\omega$. The next step is to observe that the function

$$f(\omega) = \left(\prod_{k=1}^{n} \overline{m_0(2^{k}\omega)}\, \tilde{m}_0(2^{k}\omega) \right) e^{i2^{n+1}l\omega}$$

is π-periodic, and this means that

$$\int_{-\pi/2}^{\pi/2} f(\omega)\, d\omega$$

$$= \int_{-\pi/2}^{\pi/2} f(\omega) \left[\overline{m_0(\omega)}\, \tilde{m}_0(\omega) + \overline{m_0(\omega+\pi)}\, \tilde{m}_0(\omega+\pi) \right] d\omega$$

$$= \int_{-\pi}^{\pi} f(\omega) \left[\overline{m_0(\omega)}\, \tilde{m}_0(\omega) \right] d\omega.$$

Combining these relations we conclude that

$$\int_{\mathbb{R}} \overline{\hat{h}_n(\omega)}\, \hat{\tilde{h}}_n(\omega)\, e^{il\omega}\, d\omega$$

$$= 2^{n+1} \int_{-\pi}^{\pi} \left(\prod_{k=0}^{n} \overline{m_0(2^{k}\omega)}\, \tilde{m}_0(2^{k}\omega) \right) e^{i2^{n+1}l\omega}\, d\omega$$

$$= \int_{-2^{n+1}\pi}^{2^{n+1}\pi} \left(\prod_{k=1}^{n+1} \overline{m_0(2^{-k}\omega)}\, \tilde{m}_0(2^{-k}\omega) \right) e^{i2^{n+1}l\omega}\, d\omega$$

$$= \int_{\mathbb{R}} \overline{\hat{h}_{n+1}(\omega)}\, \hat{\tilde{h}}_{n+1}(\omega)\, e^{il\omega}\, d\omega,$$

where the second equality is derived by replacing $2^{n+1}\omega$ by ω. Thus, by induction, the functions h_n and $\tilde{h}_n$ satisfy

$$\langle h_n(x-k), \tilde{h}_n(x-k') \rangle = \delta_{k,k'} \tag{4.71}$$

for all n. (Note that this argument is true for $n = 0$, in which case the product must be replaced by $\widehat{h_0}(\omega)\, \widetilde{h}_0(\omega) = 1$.)

At this point we refer to Proposition 3.4 and its proof. The hypothesis of Proposition 3.4 is exactly what we have assumed about m_0 and $\tilde{m}_0$ here, namely, that $N - b > \frac{1}{2}$ and $\tilde{N} - \tilde{b} > \frac{1}{2}$. From the proof of Proposition 3.4 we conclude that

$$|\hat{h}_n(\omega)| + |\widehat{\tilde{h}}_n(\omega)| \leq C(1 + |\omega|)^{-\frac{1}{2} - \varepsilon} \tag{4.72}$$

where C and ε do not depend on n. This assures us that $\|\hat{h}_n\|$ and $\|\widehat{\tilde{h}}_n\|$ are bounded uniformly in n. On the other hand, it is clear $\hat{h}_n(\omega)$ and $\widehat{\tilde{h}}_n(\omega)$ tend respectively to $\hat{\varphi}(\omega)$ and $\widehat{\tilde{\varphi}}(\omega)$ for each ω. One can thus apply the dominated convergence theorem and conclude that the sequences h_n and $\tilde{h}_n$ tend to φ and $\tilde{\varphi}$ in $L^2(\mathbb{R})$. Consequently, the functions $\{\varphi(x - k)\}_{k \in \mathbb{Z}}$ and $\{\tilde{\varphi}(x - k)\}_{k \in \mathbb{Z}}$ satisfy (4.68) and form a biorthogonal system. This proves Proposition 4.6. $\square$

REMARKS

Applying the Schwarz inequality to the summation formula (4.55) shows that

$$1 \leq \left[\sum_{k \in \mathbb{Z}} |\hat{\varphi}(\omega + 2k\pi)|^2\right] \left[\sum_{k \in \mathbb{Z}} |\widehat{\tilde{\varphi}}(\omega + 2k\pi)|^2\right] \tag{4.73}$$

for all ω. From the hypotheses of Proposition 4.1, this means that there are positive constants C_1, C_2, $\tilde{C}_1$, and $\tilde{C}_2$ such that the trigonometric polynomials α and $\tilde{\alpha}$ satisfy

$$0 < C_1 \leq \alpha(\omega) \leq C_2 \quad \text{and} \quad 0 < \tilde{C}_1 \leq \tilde{\alpha}(\omega) \leq C_2 \tag{4.74}$$

for all ω. This, in turn, implies that the families $\{\varphi(x - k)\}_{k \in \mathbb{Z}}$ and $\{\tilde{\varphi}(x - k)\}_{k \in \mathbb{Z}}$ are unconditional bases for the subspaces they generate. To see this, let $f(x) = \sum_{k \in \mathbb{Z}} e_k\, \varphi(x - k)$ where $\{e_k\}$ is in $l^2(\mathbb{Z})$ and write $E(\omega) = \sum_{k \in \mathbb{Z}} e_k\, e^{-ik\omega}$. Then

$$\|f\|^2 = \frac{1}{2\pi} \int_0^{2\pi} |E(\omega)|^2 \alpha(\omega)\, d\omega,$$

and it is clear that

$$C_1 \sum |e_k|^2 \leq \|f\|^2 \leq C_2 \sum |e_k|^2.$$

The same computation works for $\{\tilde{\varphi}(x - k)\}_{k \in \mathbb{Z}}$, and a similar

UNCONDITIONAL BIORTHOGONAL WAVELET BASES   121

computation shows that, at each scale, the families $\{\psi_{j,k}\}_{k\in\mathbb{Z}}$ and $\{\tilde{\psi}_{j,k}\}_{k\in\mathbb{Z}}$ are Riesz bases for the subspaces W_j and $\tilde{W}_j$.

It is also possible to define a larger subspace by considering the direct sum $W = \bigoplus_{j=-J}^{J} W_j$. The family $\{\psi_{j,k}\}_{k\in\mathbb{Z}, -J\leq j\leq J}$ will again constitute an unconditional basis for W. This comes from the fact that it is possible to define the notion of 'angle' between the subspaces W_j, and that everything then works as if one were dealing with finite dimensional spaces.

On the other hand, one has every reason to fear that the Riesz constants, which are obtained from equivalence between the Hilbert norm and the sum of the squares of the coordinates, have a tendency to blow up when J tends to infinity. This is exactly what we seek to avoid.

We wish to establish an equivalence among the following quantities:

$$\|f\|^2 = \int |f|^2 = \|\sum_{j,k} f_{j,k}\,\psi_{j,k}\|^2 = \|\sum_{j,k} \tilde{f}_{j,k}\,\tilde{\psi}_{j,k}\|^2, \quad (4.75)$$

$$\sum_{j,k} |\langle f, \psi_{j,k}\rangle|^2 = \sum_{j,k} |\tilde{f}_{j,k}|^2, \quad (4.76)$$

and

$$\sum_{j,k} |\langle f, \tilde{\psi}_{j,k}\rangle|^2 = \sum_{j,k} |f_{j,k}|^2. \quad (4.77)$$

The last step in the proof of Theorem 4.1 is to produce the constants that relate these norms to each other.

Proposition 4.7 *The hypotheses of Proposition 4.1 imply that the norms* (4.75), (4.76), *and* (4.77) *are equivalent.*

Proof. Our first step will be to prove the inequality

$$\sum_{j,k\in\mathbb{Z}} |\langle f, \psi_{j,k}\rangle|^2 \leq A\|f\|^2. \quad (4.78)$$

For this, we decompose the sum (not worrying about convergence since everything in sight is positive) as

$$\sum_{j,k\in\mathbb{Z}} |\langle f, \psi_{j,k}\rangle|^2 = \sum_{j\in\mathbb{Z}}\sum_{k\in\mathbb{Z}} |\langle f, \psi_{j,k}\rangle|^2 = \sum_{j\in\mathbb{Z}} \sigma_j(f). \quad (4.79)$$

The $\sigma_j(f)$ can be written as integrals between $-\pi$ and π using the familiar summation formulas of Lemma 4.1. In this case we have

$$\begin{aligned}\sigma_j(f) &= \frac{1}{4\pi^2}\sum_{k\in\mathbb{Z}}|\langle \hat{f},\hat{\psi}_{j,k}\rangle|^2 \\ &= \frac{2^j}{4\pi^2}\sum_{k\in\mathbb{Z}}\left|\int_{\mathbb{R}}\hat{f}(2^j\omega)\,\overline{\hat{\psi}(\omega)}\,e^{ik\omega}\,d\omega\right|^2 \\ &= \frac{2^j}{2\pi}\int_{-\pi}^{\pi}\left|\sum_{l\in\mathbb{Z}}\hat{f}(2^j(\omega+2l\pi))\,\overline{\hat{\psi}(\omega+2l\pi)}\right|^2 d\omega\,.\end{aligned}$$

At this point it is necessary to make a sharper estimate than was made in Proposition 4.2 since we wish to be able to sum the $\sigma_j(f)$ terms. From (4.24) we know that $|\hat{\psi}(\omega)|\leq C(1+|\omega|)^{-\frac{1}{2}-\varepsilon}$, and from this it follows that we can find a $\rho > 0$ so that

$$\sum_{k\in\mathbb{Z}}|\hat{\psi}(\omega+2k\pi)|^{2-\rho}\leq C_1(\rho)\,. \qquad (4.80)$$

Specifically, we choose and fix ρ with $0 < \rho < \frac{4\varepsilon}{1+2\varepsilon}$. This estimate allows us to bound $\sigma_j(f)$ as follows:

$$\begin{aligned}\sigma_j(f) &\leq \frac{2^j}{2\pi}\int_{-\pi}^{\pi}\left[\sum_{l\in\mathbb{Z}}|\hat{f}(2^j(\omega+2l\pi))|\,|\hat{\psi}(\omega+2l\pi)|\right]^2 d\omega \\ &= \frac{2^j}{2\pi}\int_{-\pi}^{\pi}\left[\sum_{l\in\mathbb{Z}}|\hat{f}(2^j(\omega+2l\pi))|\,|\hat{\psi}(\omega+2l\pi)|^{\frac{\rho}{2}}\right. \\ &\qquad\qquad \left.|\hat{\psi}(\omega+2l\pi)|^{1-\frac{\rho}{2}}\right]^2 d\omega \\ &\leq \frac{2^j}{2\pi}\int_{-\pi}^{\pi}\left[\sum_{l\in\mathbb{Z}}|\hat{f}(2^j(\omega+2l\pi))|^2\,|\hat{\psi}(\omega+2l\pi)|^{\rho}\right] \\ &\qquad\qquad \left[\sum_{l\in\mathbb{Z}}|\hat{\psi}(\omega+2l\pi)|^{2-\rho}\right]d\omega \\ &\leq \frac{C_1(\rho)}{2\pi}\int_{\mathbb{R}}|\hat{f}(\omega)|^2\,|\hat{\psi}(2^{-j}\omega)|^{\rho}\,d\omega\,.\end{aligned}$$

When we sum both sides of this inequality over j we see that the term $\sum_{j\in\mathbb{Z}}|\hat{\psi}(2^{-j}\omega)|^\rho$ appears under the integral on the right. The crux of the proof is to show that this function is bounded. Indeed, this follows from the facts that $\hat{\psi}(0) = 0$ and that $\hat{\psi}$ is continuously differentiable, plus the estimate on $\hat{\psi}(\omega)$ at infinity.

So as not to obscure the line of the proof, we will leave the details of the argument until later and assume here that

$$\sum_{j \in \mathbb{Z}} |\hat{\psi}(2^{-j}\omega)|^\rho \leq C_2(\rho). \tag{4.81}$$

From this we deduce the inequality (4.78) that we wished to establish, with a value for the constant A given by

$$A = C_1(\rho) C_2(\rho). \tag{4.82}$$

The same computation with $\tilde{\psi}$ in place of ψ shows that

$$\sum_{j,k \in \mathbb{Z}} |\langle f, \tilde{\psi}_{j,k}\rangle|^2 \leq \tilde{A}\|f\|^2. \tag{4.83}$$

To establish the inverse inequalities, which will ensure the complete equivalence among the norms, we will express $\|f\|^2$ in terms of the coordinates of f in both bases. Recall that a function f can be expressed as

$$f = \lim_{N \to +\infty} \sum_{j=-N}^{N} \sum_{k \in \mathbb{Z}} \langle f, \tilde{\psi}_{j,k}\rangle \psi_{j,k}$$

or as

$$f = \lim_{N \to +\infty} \sum_{j=-N}^{N} \sum_{k \in \mathbb{Z}} \langle f, \psi_{j,k}\rangle \tilde{\psi}_{j,k}.$$

Thus one can evaluate $\|f\|^2$ by writing

$$\begin{aligned}
\|f\|^2 &= \langle f, f \rangle \\
&= \langle \lim_{N \to +\infty} \sum_{j=-N}^{N} \sum_{k \in \mathbb{Z}} \langle f, \psi_{j,k}\rangle \tilde{\psi}_{j,k}, f \rangle \\
&= \lim_{N \to +\infty} \sum_{j=-N}^{N} \sum_{k \in \mathbb{Z}} \langle f, \psi_{j,k}\rangle \overline{\langle f, \tilde{\psi}_{j,k}\rangle}.
\end{aligned}$$

Based on the inequalities (4.78) and (4.83) that we have just demonstrated, this series converges absolutely, and we have

$$\|f\| = \left[\sum_{j,k \in \mathbb{Z}} \langle f, \psi_{j,k}\rangle \overline{\langle f, \tilde{\psi}_{j,k}\rangle} \right]^{1/2} \tag{4.84}$$

and

$$\|f\|^2 \leq \left[\sum_{j,k\in\mathbb{Z}} |\langle f, \psi_{j,k}\rangle|^2\right]^{1/2} \left[\sum_{j,k\in\mathbb{Z}} |\langle f, \tilde{\psi}_{j,k}\rangle|^2\right]^{1/2}. \quad (4.85)$$

We can now obtain the required equivalences by substituting the estimates (4.78) and (4.83) in this last inequality. This gives us, respectively,

$$\frac{1}{A}\|f\|^2 \leq \sum_{j,k\in\mathbb{Z}} |\langle f, \tilde{\psi}_{j,k}\rangle|^2 \leq \tilde{A}\|f\|^2 \quad (4.86)$$

and

$$\frac{1}{\tilde{A}}\|f\|^2 \leq \sum_{j,k\in\mathbb{Z}} |\langle f, \psi_{j,k}\rangle|^2 \leq A\|f\|^2. \quad (4.87)$$

This completes the proof of Theorem 4.1, except for the details involved in establishing the inequality (4.81). Since sums of this kind are important in the multiscale culture, we include a brief demonstration in the following lemma. Our version is surely not the strongest, but it is general enough to include (4.81). □

Lemma 4.2 *Let g be defined and continuously differentiable on $\mathbb{R}$. Assume further that $g(0) = 0$ and that*

$$|g(x)| \leq C(1+|x|)^{-\delta}$$

for some $\delta > 0$. Then, for all $\rho > 0$,

$$\sum_{j\in\mathbb{Z}} |g(2^{-j}x)|^\rho \leq C(\rho),$$

where $C(\rho)$ is a constant that depends on ρ but not on x.

Proof. Write $G(x) = \sum_{j\in\mathbb{Z}} |g(2^{-j}x)|^\rho$ and observe that

$$\sup_{x\in\mathbb{R}} G(x) = \sup_{1\leq |x|\leq 2} G(x).$$

Break the sum into two parts so G is expressed as

$$G(x) = \sum_{j=0}^{\infty} |g(2^{-j}x)|^\rho + \sum_{j=1}^{\infty} |g(2^j x)|^\rho.$$

Since g' is continuous, we can define $D = \max_{|x|\leq 2} |g'(x)|$. Then, since $g(0) = 0$, we have $|g(x)| \leq D|x|$ for $|x| \leq 2$. Turning to the first sum, we have

$$\sum_{j=0}^{\infty} |g(2^{-j}x)|^\rho \leq D^\rho \sum_{j=0}^{\infty} |2^{-j}x|^\rho = D^\rho |x|^\rho \frac{1}{1-2^{-\rho}},$$

so that

$$\sum_{j=0}^{\infty} |g(2^{-j}x)|^\rho \leq \frac{(2D)^\rho}{1-2^{-\rho}} \qquad (4.88)$$

for $|x| \leq 2$.

We use the decay hypothesis to estimate the second sum. Thus,

$$\sum_{j=1}^{\infty} |g(2^j x)|^\rho \leq C^\rho \sum_{j=1}^{\infty} (1+|2^j x|)^{-\delta\rho} \leq \frac{C^\rho}{1-2^{-\delta\rho}} \qquad (4.89)$$

for $|x| \geq 1$. Combining (4.88) and (4.89) we see that

$$\sum_{j \in \mathbb{Z}} |g(2^{-j}x)|^\rho \leq \left[\frac{C^\rho}{1-2^{-\delta\rho}} + \frac{(2D)^\rho}{1-2^{-\rho}} \right] \qquad (4.90)$$

for $1 \leq |x| \leq 2$, which proves the lemma. □

REMARKS

• Inequalities (4.78) and (4.83) in no way imply that one has biorthogonality properties of the type (4.54). They do however imply that one has 'semi-frames' or 'Bessel sequences' in the sense defined in Young (1980).

• On the other hand, the biorthogonality was fully exploited in obtaining the equivalences (4.86) and (4.87) and the equality (4.84). Note that this equality tells us that the quadratic norm of f can be estimated precisely using the pyramid algorithms, provided that one wishes to take the trouble to do the two decompositions. In any case, the equivalences ensure the stability of both decompositions as well as that of the reconstructions.

• Our estimates of the constants A and $\tilde{A}$ may seem quite rough in view of the dependence on the parameter ρ. It would be nice to choose this parameter so as to optimize the equivalences, which means to have the product $A\tilde{A}$ as close as possible to 1. Our desires are not ill-founded. When the functions m_0 and $\tilde{m}_0$ provide good frequency separation, which means that they approach the degenerate case where $m_0 = \tilde{m}_0 = \mathbf{1}_{[-\pi/2,\,\pi/2]}$, then the estimates of the constants given by (4.82) get better. In fact, for the degenerate case, it is not difficult to show that $A = \tilde{A} = 1$, independently of the value of ρ. Thus we may hope that the quadratic estimates involve 'reasonable' constants in most cases.

- It is clearly possible to construct biorthogonal bases for $l^2(\mathbb{Z})$ by proceeding along the lines presented in Chapter 1 for the orthogonal case. We will not present this construction here because we believe that the connections between the discrete and continuous points of view have been well established in the preceding chapters.

- The fact that the Riesz constants do not explode when the subspaces W_j stack up can be interpreted as follows: These spaces are 'asymptotically' orthogonal in the sense that, if we have a sequence of functions $\{f_j\}_{j \in \mathbb{Z}}$ of W_j such that $\|f_j\| = 1$, then $\lim_{(p-q) \to +\infty} \langle f_p, f_q \rangle = 0$. This results from $\hat{\psi}(2^{-p}\omega)$ and $\hat{\psi}(2^{-q}\omega)$ being concentrated on dyadic intervals that become well separated as $(p - q) \to +\infty$.

- Finally, we note again that the assumptions we have made in this section about m_0 and $\tilde{m}_0$, which imply the decay at infinity of φ and $\tilde{\varphi}$ given by (4.24), are in no sense minimal; they are not strictly necessary to obtain unconditional biorthogonal wavelet bases. These assumptions are, however, in the same spirit as our spectral analysis of regularity. In the next section we take a different point of view and establish necessary and sufficient conditions on the filters m_0 and $\tilde{m}_0$ so that they generate unconditional biorthogonal wavelet bases. For this, we return to the transition operators introduced in Chapter 2.

4.4 Dual filters and biorthogonal Riesz bases

The basic assumptions for this section are the same as before, namely, equations (4.14) through (4.18). We also assume that the filters m_0 and $\tilde{m}_0$ are trigonometric polynomials and are 'low-pass' in the sense that

$$m_0(0) = \tilde{m}_0(0) = 1 \quad \text{and} \quad m_0(\pi) = \tilde{m}_0(\pi) = 0 \ . \quad (4.91)$$

However, in this section, we no longer assume that m_0 and $\tilde{m}_0$ satisfy the hypotheses of Proposition 4.1.

In this section, which is based on joint work by Albert Cohen and Ingrid Daubechies (1992), we will present necessary and sufficient conditions on the filters m_0 and $\tilde{m}_0$ so that $\{\psi_{j,k}\}_{j,k \in \mathbb{Z}}$ and $\{\tilde{\psi}_{j,k}\}_{j,k \in \mathbb{Z}}$ form unconditional biorthogonal wavelet bases. These conditions will be given in terms of the eigenvalues of the transition operators P_0 and $\tilde{P}_0$. But before broaching this result, we encourage the reader to review the proofs of Propositions 4.2, 4.3, 4.4, and 4.5 to become convinced that the conclusions of these propositions hold if we assume that $\hat{\varphi}$ and $\hat{\tilde{\varphi}}$ are in $L^2(\mathbb{R})$. With these assumptions, which are weaker than the hypotheses of Proposition 4.1, some of the equations in these proofs must be interpreted as holding almost everywhere or as elements of $L^2(\mathbb{R})$. This is particularly true for those expressions that use a form of the Poisson summation formula, which in this case are based on parts i), ii), and iii) of Lemma 4.1. In the same spirit, we note that Proposition 4.6 holds if we also know that the truncated products $\hat{h}_n$ and $\hat{\tilde{h}}_n$ tend to $\hat{\varphi}$ and $\hat{\tilde{\varphi}}$ in $L^2(\mathbb{R})$. The situation for Proposition 4.7, and hence for the completion of Theorem 4.1, is more complicated. For this, we need to have the estimates (4.80) and (4.81). Developing these estimates without the assumptions of Proposition 4.1 will be an important aspect of this section.

To begin our program, we reintroduce the transition operators P_0 and $\tilde{P}_0$ associated with m_0 and $\tilde{m}_0$ that act on the 2π-periodic functions. They are defined by

$$P_0 f(\omega) = |m_0(\frac{\omega}{2})|^2 f(\frac{\omega}{2}) + |m_0(\frac{\omega}{2} + \pi)|^2 f(\frac{\omega}{2} + \pi) \qquad (4.92)$$

and

$$\tilde{P}_0 f(\omega) = |\tilde{m}_0(\frac{\omega}{2})|^2 f(\frac{\omega}{2}) + |\tilde{m}_0(\frac{\omega}{2} + \pi)|^2 f(\frac{\omega}{2} + \pi). \qquad (4.93)$$

Recall from Section 2.4 that these operators can be studied in finite dimensional spaces. In particular, if the filter m_0 is written as $m_0(\omega) = \sum_{k=N_1}^{N_2} h_k e^{-ik\omega}$, then, with $N = N_2 - N_1$, the $(2N+1)$-dimensional subspace E_N defined by

$$E_N = \left\{ \sum_{k=-N}^{N} c_k e^{ik\omega} \,\middle|\, (c_{-N}, \ldots, c_N) \in \mathbb{C}^{2N+1} \right\}$$

is stable under the action of P_0, that is, $P_0(E_N) \subset E_N$.

Define the subspace $F_N \subset E_N$ by

$$F_N = \left\{ \sum_{k=-N}^{N} c_k e^{ik\omega} \,\middle|\, \sum_{k=-N}^{N} c_k = 0 \right\}.$$

A trigonometric polynomial f is in F_N if and only if $f \in E_N$ and $f(0) = 0$. Hence, if $f \in F_N$, we have

$$P_0 f(0) = |m_0(0)|^2 f(0) + |m_0(\pi)|^2 f(\pi) = 0,$$

which shows that F_N is also stable under the action of P_0.

The following result is due to Jean-Pierre Conze. It is a key element in our program because it links the iterates of the operator P_0 to the truncated products $\hat{h}_n$.

Lemma 4.3 *If f is any 2π-periodic, continuous function, then*

$$\int_{-\pi}^{\pi} (P_0^n f)(\omega)\, d\omega = \int_{-2^n\pi}^{2^n\pi} \left[\prod_{k=1}^{n} |m_0(2^{-k}\omega)|^2\right] f(2^{-n}\omega)\, d\omega \quad (4.94)$$

for all $n \geq 1$.

Proof. The proof proceeds by induction. For $n = 1$ we have

$$\int_{-\pi}^{\pi}(P_0 f)(\omega)\,d\omega = \int_{-\pi}^{\pi}\left[|m_0(\frac{\omega}{2})|^2 f(\frac{\omega}{2}) + |m_0(\frac{\omega}{2}+\pi)|^2 f(\frac{\omega}{2}+\pi)\right]d\omega$$

$$= \int_{-2\pi}^{2\pi} |m_0(2^{-1}\omega)|^2 f(2^{-1}\omega)\,d\omega\,.$$

Now assume that the result has been verified through $n-1$. Then

$$\int_{-\pi}^{\pi}(P_0^n f)(\omega)\,d\omega = \int_{-\pi}^{\pi}(P_0^{n-1}(P_0 f))(\omega)\,d\omega$$

$$= \int_{-2^{n-1}\pi}^{2^{n-1}\pi}\left[\prod_{k=1}^{n-1}|m_0(2^{-k}\omega)|^2\right](P_0 f)(2^{-n+1}\omega)\,d\omega$$

$$= \int_{-2^{n-1}\pi}^{2^{n-1}\pi}\left[\prod_{k=1}^{n-1}|m_0(2^{-k}\omega)|^2\right]$$
$$\left[|m_0(2^{-n}\omega)|^2 f(2^{-n}\omega)\right.$$
$$\left.+|m_0(2^{-n}(\omega+2^n\pi))|^2 f(2^{-n}(\omega+2^n\pi))\right]d\omega$$

$$= \int_{-2^n\pi}^{2^n\pi}\left[\prod_{k=1}^{n}|m_0(2^{-k}\omega)|^2\right]f(2^{-n}\omega)\,d\omega\,.$$

By induction, the result holds for all $n \geq 1$. □

The connection with $\hat{h}_n$ is immediate:

$$\int_{-\pi}^{\pi}(P_0^n f)(\omega)\,d\omega = \int_{\mathbb{R}}|\hat{h}_n(\omega)|^2 f(2^{-n}\omega)\,d\omega \quad (4.95)$$

for any 2π-periodic, continuous function f. This shows that it should be possible to estimate the L^2 norms of the $\hat{h}_n$ in terms of iterates of P_0 by making an appropriate choice for the function f. The following proposition, which incorporates this idea, constitutes essentially the 'sufficient' half of our main result.

DUAL FILTERS AND BIORTHOGONAL RIESZ BASES   129

Proposition 4.8 *Let λ be the largest eigenvalue of P_0 restricted to F_N. If $|\lambda| < 1$ then h_n converges to φ in $L^2(\mathbb{R})$.*

Proof. We will show that $\hat{h}_n$ tends to $\hat{\varphi}$ in $L^2(\mathbb{R})$. The result follows by taking Fourier transforms.

Our choice for f is the trigonometric polynomial c defined by

$$c(\omega) = 1 - \cos \omega. \tag{4.96}$$

It is clear that $c \in F_N$. (We assume that $N \geq 1$.) Furthermore, $c(\omega) \geq 0$, and $c(\omega) \geq 1$ when $\pi/2 \leq |\omega| \leq \pi$. Applying Lemma 4.3, we see that

$$\begin{aligned}
\int |\hat{h}_n(\omega)|^2 c(2^{-n}\omega) \, d\omega &= \int_{-\pi}^{\pi} (P_0^n c)(\omega) \, d\omega \\
&\leq \sqrt{2\pi} \|P_0^n c\| \\
&\leq K \rho^n
\end{aligned}$$

whenever $\rho > |\lambda|$. ($K = \sqrt{2\pi} \|c\| = \sqrt{6}\pi$.)

The next step relates $\hat{\varphi}(\omega)$ to $\hat{h}_n(\omega)$ when $|\omega| \leq 2^n \pi$. For this range, we have $\hat{\varphi}(\omega) = \hat{h}_n(\omega) \, \hat{\varphi}(2^{-n}\omega)$, which yields the estimate

$$|\hat{\varphi}(\omega)| \leq M_{\hat{\varphi}} |\hat{h}_n(\omega)| \quad \text{for all} \quad |\omega| \leq 2^n \pi, \tag{4.97}$$

where $M_{\hat{\varphi}} = \max_{|\omega| \leq \pi} |\hat{\varphi}(\omega)|$.

We are now ready to estimate the L^2 norm of $\hat{\varphi}$. We do this by estimating the norms of the Littlewood–Paley blocks of φ. Thus, for $n \geq 1$, define the operator $\hat{\Delta}_n$ by

$$\hat{\Delta}_n \varphi(\omega) = \hat{\varphi}(\omega) \left[\mathbf{1}_{[-2^n \pi, \, -2^{n-1} \pi]}(\omega) + \mathbf{1}_{[2^{n-1} \pi, \, 2^n \pi]}(\omega) \right].$$

For $n = 0$, define $\hat{\Delta}_0 \varphi(\omega) = \hat{\varphi}(\omega) \mathbf{1}_{[-\pi, \, \pi]}(\omega)$. With this notation, $\|\hat{\varphi}\|^2 = \sum_{n=0}^{\infty} \|\hat{\Delta}_n \varphi\|^2$. We estimate $\|\hat{\Delta}_n \varphi\|^2$ as follows:

$$\begin{aligned}
\|\hat{\Delta}_n \varphi\|^2 &= \int_{2^{n-1}\pi \leq |\omega| \leq 2^n \pi} |\hat{\varphi}(\omega)|^2 \, d\omega \\
&\leq M_{\hat{\varphi}}^2 \int_{2^{n-1}\pi \leq |\omega| \leq 2^n \pi} |\hat{h}_n(\omega)|^2 \, d\omega \\
&\leq M_{\hat{\varphi}}^2 \int_{-2^n \pi}^{2^n \pi} |\hat{h}_n(\omega)|^2 c(2^{-n}\omega) \, d\omega \\
&\leq M_{\hat{\varphi}}^2 K \rho^n.
\end{aligned}$$

This inequality holds for all $n \geq 0$ and for all $\rho > |\lambda|$. Since we have assumed that $|\lambda| < 1$, we can choose ρ so that $|\lambda| < \rho < 1$.

With this choice for ρ, it is clear that the series $\sum_{n=0}^{\infty} \|\hat{\Delta}_n \varphi\|^2$ converges and that $\hat{\varphi}$ is in $L^2(\mathbb{R})$.

The geometric decay of these Littlewood–Paley blocks plays a central role in several proofs. Thus, for future reference, we write $C = M_{\hat{\varphi}}^2 K$, record the estimate

$$\|\hat{\Delta}_n \varphi\|^2 \leq C\rho^n, \tag{4.98}$$

and mention specifically that C depends only on φ.

We use this result in the first place to prove that $\hat{h}_n$ converges to $\hat{\varphi}$ in $L^2(\mathbb{R})$. Since $\hat{\varphi}(0) = 1$ and $\hat{\varphi}$ is regular, there exists an α in $]0, \pi]$ such that

$$|\hat{\varphi}(\omega)| \geq \frac{1}{2}$$

when $|\omega| \leq \alpha$. We divide $\hat{h}_n$ into two parts, $\hat{h}_n = \hat{h}_{n,1} + \hat{h}_{n,2}$, by defining

$$\hat{h}_{n,1}(\omega) = \hat{h}_n(\omega) \mathbf{1}_{[-2^n \alpha, 2^n \alpha]}(\omega),$$

and

$$\hat{h}_{n,2}(\omega) = \hat{h}_n(\omega) \left[\mathbf{1}_{[-2^n \pi, 2^n \pi]}(\omega) - \mathbf{1}_{[-2^n \alpha, 2^n \alpha]}(\omega) \right].$$

It is clear that $\hat{h}_{n,1}(\omega)$ tends to $\hat{\varphi}(\omega)$ for all ω. Furthermore, since

$$\hat{\varphi}(\omega) = \hat{h}_{n,1}(\omega)\hat{\varphi}(2^{-n}\omega)$$

for $|\omega| \leq 2^n \alpha$, and since $\hat{h}_{n,1}(\omega) = 0$ for $|\omega| > 2^n \alpha$, we have

$$|\hat{h}_{n,1}(\omega)| \leq 2|\hat{\varphi}(\omega)| \tag{4.99}$$

for all ω. Since we already know that $\hat{\varphi}$ is in $L^2(\mathbb{R})$, this implies by dominated convergence that $\hat{h}_{n,1}$ tends to $\hat{\varphi}$ in $L^2(\mathbb{R})$. For the function $\hat{h}_{n,2}$ we have

$$\begin{aligned}
\int |\hat{h}_{n,2}(\omega)|^2 \, d\omega &= \int_{2^n \alpha \leq |\omega| \leq 2^n \pi} |\hat{h}_n(\omega)|^2 \, d\omega \\
&\leq \frac{1}{c(\alpha)} \int |\hat{h}_n(\omega)|^2 c(2^{-n}\omega) \, d\omega \\
&\leq \frac{K}{c(\alpha)} \rho^n
\end{aligned}$$

whenever $\rho > |\lambda|$. By taking $\rho < 1$, we see that $\hat{h}_{n,2}$ tends to zero in $L^2(\mathbb{R})$ and conclude that $\hat{\varphi}$ is the L^2 limit of the sequence $\hat{h}_n$. $\square$

REMARKS
From (4.98) we see that

$$\sum_{j \geq 0} 2^{2sj} \|\hat{\Delta}_j \varphi\|^2 < +\infty$$

whenever $s < -\frac{1}{2} \log |\lambda| / \log 2$. This means that φ is in the Sobolev space $H^s(\mathbb{R})$, that is, $\int (1+|\omega|^2)^s |\hat{\varphi}(\omega)|^2 \, d\omega < +\infty$, for these values of s. It is thus possible to use the transition operator P_0 to estimate the regularity of φ in the sense of the exponent $s_2(\varphi)$ that we have defined in Chapter 3.

Furthermore, if m_0 has a higher-order zero at π, that is, if $\left(\frac{d}{d\omega}\right)^k m_0(\pi) = 0$ for $k = 0, \ldots, N-1$, then P_0 leaves invariant the subspace $G_N = \left\{ f \in E_N \mid \left(\frac{d}{d\omega}\right)^k f(0) = 0 \text{ for } k = 0, \ldots, 2N-1 \right\}$. One can then replace λ by the largest eigenvalue of P_0 restricted to G_N to obtain a sharper regularity estimate, and, indeed, one obtains a sharp estimate of $s_2(\varphi)$ with this strategy (see Cohen and Daubechies (1993), Villemoes (1992), Eirola (1992)).

Proposition 4.8, taken in view of the comments at the beginning of this section, shows that we recover many of the results of Section 4.3 whenever the largest eigenvalues of P_0 and $\tilde{P}_0$, restricted to F_N and $\tilde{F}_N$ respectively, both have absolute value less than 1. Specifically, if we assume that these eigenvalues have absolute value less than 1, then Proposition 4.8 states that φ and $\hat{\varphi}$ are in $L^2(\mathbb{R})$ and that they are the L^2 limits of the sequences h_n and $\tilde{h}_n$. This, in turn, implies the conclusions of Propositions 4.2 through 4.6. As mentioned before, we cannot yet recover the conclusion of Theorem 4.1 because we lack the result of Proposition 4.7. To recover the conclusion of Proposition 4.7, which is that the norms (4.75), (4.76), and (4.77) are equivalent, we need the estimates (4.80) and (4.81). Once we have these estimates, the proof of Proposition 4.7 follows, and we recover the conclusion of Theorem 4.1. These estimates are the subject of the next result, which we state as a lemma. The proof depends on the geometric decay of the Littlewood–Paley blocks, (4.98)

Lemma 4.4 *Let λ be the largest eigenvalue of P_0 restricted to F_N. If $|\lambda| < 1$, then there exists $\varepsilon, \sigma > 0$ and two constants $C_1(\varepsilon)$ and $C_2(\sigma)$, which depend only on ε, σ, and φ, such that for all ω*

$$\sum_{k \in \mathbb{Z}} |\hat{\psi}(\omega + 2k\pi)|^{2-\varepsilon} \leq C_1(\varepsilon) \qquad (4.100)$$

and
$$\sum_{j\in\mathbb{Z}} |\hat{\psi}(2^{-j}\omega)|^\sigma \leq C_2(\sigma). \tag{4.101}$$

Proof. We begin with several observations about $\hat{\varphi}$ and its derivatives. Since φ has compact support, $\hat{\varphi}$ has continuous derivatives, and this implies that the function $|\hat{\varphi}|^{2-\varepsilon}$ has a continuous derivative when $0 \leq \varepsilon < 1$. To see this, first note that

$$\frac{d}{d\omega}|\hat{\varphi}|^{2-\varepsilon}(\omega) = \frac{d}{d\omega}\big[|\hat{\varphi}|^2\big]^{1-\frac{\varepsilon}{2}}(\omega) = \Big(1-\frac{\varepsilon}{2}\Big)\frac{d}{d\omega}|\hat{\varphi}|^2(\omega)\,|\hat{\varphi}(\omega)|^{-\varepsilon}.$$

When we write the derivative of $|\hat{\varphi}|^2 = \hat{\varphi}\overline{\hat{\varphi}}$ as

$$\frac{d}{d\omega}|\hat{\varphi}|^2(\omega) = \hat{\varphi}(\omega)\overline{\frac{d\hat{\varphi}}{d\omega}(\omega)} + \overline{\hat{\varphi}(\omega)}\frac{d\hat{\varphi}}{d\omega}(\omega)$$

it becomes clear that the right-hand side of the first equation is continuous if $0 \leq \varepsilon < 1$. Knowing that this derivative is continuous allows us to write

$$|\hat{\varphi}(\omega_2)|^{2-\varepsilon} - |\hat{\varphi}(\omega_1)|^{2-\varepsilon} = \int_{\omega_1}^{\omega_2} \frac{d}{d\omega}|\hat{\varphi}|^{2-\varepsilon}(\omega)\,d\omega$$

for all ω_1 and ω_2.

These observations and the fact that $\hat{\varphi}(2k\pi) = \delta_{0,k}$ allow us to estimate the value of $|\hat{\varphi}(\omega)|^2$. Thus, for $n \geq 1$ and for ω in the dyadic ring I_n defined by $2^{n-1}\pi \leq |\omega| \leq 2^n\pi$, we have

$$\begin{aligned}
|\hat{\varphi}(\omega)|^2 &\leq \int_{I_n} \left|\frac{d}{d\omega}|\hat{\varphi}|^2(\omega)\right| d\omega \\
&\leq 2\int_{I_n} |\hat{\varphi}(\omega)|\left|\frac{d\hat{\varphi}}{d\omega}(\omega)\right| d\omega \\
&\leq 2\left(\int_{\mathbb{R}} \left|\frac{d\hat{\varphi}}{d\omega}(\omega)\right|^2 d\omega\right)^{1/2} \left(\int_{I_n} |\hat{\varphi}(\omega)|^2\, d\omega\right)^{1/2}.
\end{aligned}$$

The first integral in the last line is equal to $\sqrt{2\pi}$ times the L^2 norm of $x\varphi(x)$, which is finite because φ has compact support. Thus $|\hat{\varphi}(\omega)|^2$ is bounded by a constant times $\|\hat{\Delta}_n\varphi\|$ for $n \geq 1$, and we conclude from (4.98) that, for $2^{n-1}\pi \leq |\omega| \leq 2^n\pi$ and $n \geq 1$,

$$|\hat{\varphi}(\omega)|^2 \leq C\rho^{n/2} \tag{4.102}$$

whenever $|\lambda| < \rho$. The constant C depends only on φ. By manipulating the various quantities involved, (4.102) shows that

$$|\hat{\varphi}(\omega)| \leq C^{1/2}\pi^{-\frac{\log|\lambda|}{4\log 2}}|\omega|^{\frac{\log\rho}{4\log 2}}.$$

for all $|\omega| \geq \pi$. Because $\hat{\varphi}$ is continuous, and thus bounded on $[-\pi, \pi]$, we conclude, after a suitable change of constant, that

$$|\hat{\varphi}(\omega)| \leq C'(1+|\omega|)^{-\varepsilon} \tag{4.103}$$

for all ω, with, for example, $\varepsilon = -\frac{\log|\frac{1+\lambda}{2}|}{4\log 2} > 0$.

In view of (4.17), $\hat{\psi}$ also satisfies (4.103), with a possible change of constant. This and the fact that $\hat{\psi}(0) = 0$ show that $\hat{\psi}$ satisfies the hypotheses of Lemma 4.2. Thus, by Lemma 4.2, (4.101) holds for every $\sigma > 0$.

We use the same techniques to prove (4.100). Indeed, the following estimates should be clear from what has already been said:

$$\sum_{k \in \mathbb{Z}} |\hat{\varphi}(\omega + 2k\pi)|^{2-\varepsilon} \leq \int \left|\frac{d}{d\omega}|\hat{\varphi}|^{2-\varepsilon}(\omega)\right| d\omega$$

$$= \int \left|\frac{d}{d\omega}(|\hat{\varphi}|^2)^{1-\frac{\varepsilon}{2}}(\omega)\right| d\omega$$

$$= \int \left(1-\frac{\varepsilon}{2}\right)\left|\frac{d}{d\omega}|\hat{\varphi}|^2(\omega)\right| |\hat{\varphi}(\omega)|^{-\varepsilon} d\omega$$

$$\leq (2-\varepsilon) \int \left|\frac{d\hat{\varphi}}{d\omega}(\omega)\right| |\hat{\varphi}(\omega)|^{1-\varepsilon} d\omega$$

$$\leq 2 \left(\int \left|\frac{d\hat{\varphi}}{d\omega}(\omega)\right|^2 d\omega \right)^{1/2} \left(\int |\hat{\varphi}(\omega)|^{2-2\varepsilon} d\omega \right)^{1/2}.$$

Thus, we have

$$\sum_{k \in \mathbb{Z}} |\hat{\varphi}(\omega + 2k\pi)|^{2-\varepsilon} \leq C \left(\int |\hat{\varphi}(\omega)|^{2-2\varepsilon} d\omega \right)^{1/2}, \tag{4.104}$$

and the constant C depends only on φ.

We estimate this last integral by restricting the range of integration to dyadic rings and estimating their individual contributions. By applying Hölder's inequality to these integrals (with $1/p = 1-\varepsilon$ and $1/q = \varepsilon$), we see that

$$\int_{I_n} |\hat{\varphi}(\omega)|^{2-2\varepsilon} d\omega \leq (2^n \pi)^{\varepsilon} \left(\int_{I_n} |\hat{\varphi}(\omega)|^2 d\omega \right)^{1-\varepsilon},$$

where we have again denoted the dyadic ring $2^{n-1}\pi \leq |\omega| \leq 2^n \pi$

by I_n. If we now use (4.98), with $|\lambda| < \rho < 1$, we have

$$\int_{I_n} |\hat{\varphi}(\omega)|^{2-2\varepsilon} d\omega \le C^{1-\varepsilon}\pi^{\varepsilon}(2^{\varepsilon}|\rho|^{1-\varepsilon})^n, \qquad (4.105)$$

where C depends only on φ. Thus $\sum \int_{I_n} |\hat{\varphi}(\omega)|^{2-2\varepsilon} d\omega$ will converge and the integral in (4.104) will be finite if we can choose ε so that $2^{\varepsilon}|\rho|^{1-\varepsilon} < 1$. But for this we only need to choose $\varepsilon < \frac{-\log|\rho|}{-\log|\rho|+\log 2}$. We are then assured that $0 < \varepsilon < 1$ since $|\rho| < 1$. This completes the proof of the lemma.

As the proof shows, (4.101) is true for all σ. On the other hand, for (4.100) to hold, it is necessary to choose $0 < \varepsilon < \frac{-\log|\lambda|}{-\log|\lambda|+\log 2}$.
$\square$

As mentioned above, Lemma 4.4 and the other results of this section allow us to recover the conclusions of Section 4.3 under hypotheses that are weaker than those given in Proposition 4.1. More precisely, the results of this section imply that the families $\{\psi_{j,k}\}$ and $\{\tilde{\psi}_{j,k}\}$ form unconditional biorthogonal bases whenever the largest eigenvalues of P_0 and $\tilde{P}_0$, restricted to F_N and $\tilde{F}_N$ respectively, both have absolute value less than 1. The converse is also true, and together these results form the main theorem of this section. Because the converse is slightly more technical, we will state and prove the main theorem after we have developed some preliminary results.

As a first step toward establishing the converse, we examine the consequences for the transition operator P_0 of the assumption '$\{\varphi(x-k)\}_{k\in\mathbb{Z}}$ is a Riesz basis.'

We have seen in Chapter 1 (Section 1.2) that $\{\varphi(x-k)\}_{k\in\mathbb{Z}}$ being a Riesz basis implies the existence of two positive constants C_1 and C_2, $0 \le C_1 < C_2$, such that

$$C_1 \le \sum_{l\in\mathbb{Z}} |\hat{\varphi}(\omega+2l\pi)|^2 \le C_2. \qquad (4.106)$$

Recall that this sum is equal almost everywhere to a trigonometric polynomial whose coefficients are given by $\langle \varphi(x), \varphi(x-k) \rangle$ (Lemma 4.1, iii)). We denote this function by m.

It was also shown in Section 1.2 that the family $\{\phi(x-k)\}_{k\in\mathbb{Z}}$, where ϕ is defined by

$$\hat{\phi}(\omega) = \hat{\varphi}(\omega)\Big(\sum_{l\in\mathbb{Z}} |\hat{\varphi}(\omega+2l\pi)|^2\Big)^{-1/2}, \qquad (4.107)$$

is an orthonormal basis. In the time domain we have

$$\phi(x) = \sum_{k \in \mathbb{Z}} \alpha_k \varphi(x - k) \tag{4.108}$$

where α_k is the coefficient of $e^{-ik\omega}$ in the Fourier decomposition of the 2π-periodic function $(\sum_{l \in \mathbb{Z}} |\hat{\varphi}(\omega + 2l\pi)|^2)^{-1/2} = (m(\omega))^{-1/2}$. Since the trigonometric polynomial m is bounded away from zero by (4.106), $(m(\omega))^{-1/2}$ is a smooth 2π-periodic function whose Fourier coefficients α_k decay rapidly at infinity.

The definition of ϕ implies that

$$\sum_{l \in \mathbb{Z}} |\hat{\phi}(\omega + 2l\pi)|^2 = 1, \tag{4.109}$$

which expresses the orthogonality of the translates of ϕ. If the scaling function φ is defined by (4.15), then

$$\hat{\phi}(\omega) = \prod_{k=1}^{+\infty} p_0(2^{-k}\omega) \tag{4.110}$$

where

$$p_0(\omega) = m_0(\omega) \left[\frac{m(2\omega)}{m(\omega)} \right]^{-1/2}. \tag{4.111}$$

In general, p_0 is not a trigonometric polynomial, but it is a $\mathcal{C}^\infty$, 2π-periodic function. Furthermore, a quick computation, using the equation $\hat{\varphi}(2\omega) = m_0(\omega)\hat{\varphi}(\omega)$, shows that p_0 satisfies the CQF relations

$$|p_0(\omega)|^2 + |p_0(\omega + \pi)|^2 = 1, \tag{4.112}$$

and

$$|p_0(0)| = 1, \tag{4.113}$$

which are characteristics of the orthonormal case.

Let P_0 and R_0 be the transition operators associated, respectively, with m_0 and p_0. The following lemma relates the spectral properties of these operators.

Lemma 4.5 *Assume that λ is an eigenvalue of P_0 restricted to E_N and that f_λ is a trigonometric polynomial in E_N such that $P_0 f_\lambda = \lambda f_\lambda$. Then λ is also an eigenvalue of R_0 and $R_0 g_\lambda = \lambda g_\lambda$, where g_λ is given by*

$$g_\lambda(\omega) = \frac{f_\lambda(\omega)}{m(\omega)}. \tag{4.114}$$

Proof. The proof amounts to observing that, if $g(\omega) = \frac{f(\omega)}{m(\omega)}$ for $f \in E_N$, then

$$R_0 g(\omega) = \left|p_0\left(\frac{\omega}{2}\right)\right|^2 g\left(\frac{\omega}{2}\right) + \left|p_0\left(\frac{\omega}{2}+\pi\right)\right|^2 g\left(\frac{\omega}{2}+\pi\right)$$

$$= \left|m_0\left(\frac{\omega}{2}\right)\right|^2 \frac{f\left(\frac{\omega}{2}\right)}{m(\omega)} + \left|m_0\left(\frac{\omega}{2}+\pi\right)\right|^2 \frac{f\left(\frac{\omega}{2}+\pi\right)}{m(\omega)}$$

$$= \frac{P_0 f(\omega)}{m(\omega)},$$

and to applying this to g_λ. □

Note that g_λ is not a trigonometric polynomial; it is, however, the ratio of two trigonometric polynomials. We will use this property in the next proposition where we use the eigenfunction g_λ of R_0 that corresponds to an eigenvalue λ of P_0 to derive information about λ.

Proposition 4.9 *If $\{\varphi(x-k)\}_{k\in\mathbb{Z}}$ is a Riesz basis and if λ is the largest eigenvalue of P_0 restricted to F_N, then $|\lambda| < 1$.*

Proof. We see from the computation in Lemma 4.5 that our task is equivalent to studying the eigenvalues of R_0 restricted to the set $G_N = \{f/m \mid f \in E_N\}$.

As a first step, we prove that $|\lambda| \leq 1$. For this, let g_λ be the eigenvector of R_0 associated with λ as in Lemma 4.5 and choose ω_0 such that

$$|g_\lambda(\omega_0)| = \max_{\omega \in \mathbb{R}} |g_\lambda(\omega)| > 0. \qquad (4.115)$$

Then

$$|\lambda g_\lambda(\omega_0)| = |R_0 g_\lambda(\omega_0)|$$

$$= \left|\left|p_0\left(\frac{\omega_0}{2}\right)\right|^2 g_\lambda\left(\frac{\omega_0}{2}\right) + \left|p_0\left(\frac{\omega_0}{2}+\pi\right)\right|^2 g_\lambda\left(\frac{\omega_0}{2}+\pi\right)\right|$$

$$\leq \left|p_0\left(\frac{\omega_0}{2}\right)\right|^2 \left|g_\lambda\left(\frac{\omega_0}{2}\right)\right| + \left|p_0\left(\frac{\omega_0}{2}+\pi\right)\right|^2 \left|g_\lambda\left(\frac{\omega_0}{2}+\pi\right)\right|.$$

Since $|g_\lambda(\omega_0)|$ is the maximum value of $|g_\lambda(\omega)|$, we conclude that

$$|\lambda g_\lambda(\omega_0)| \leq \left(\left|p_0\left(\frac{\omega_0}{2}\right)\right|^2 + \left|p_0\left(\frac{\omega_0}{2}+\pi\right)\right|^2\right) |g_\lambda(\omega_0)|$$

$$= |g_\lambda(\omega_0)|,$$

and this implies that $|\lambda| \leq 1$.

The next step, which is the crux, is to show that λ cannot lie on

DUAL FILTERS AND BIORTHOGONAL RIESZ BASES 137

the unit circle. This can be derived directly by noting that R_0 leaves invariant the positive cone of G_N, which consists of the $g \in G_N$ such that $g(\omega) \geq 0$. One can then use the Krein–Rutman theorem (see Schaefer (1974) for a full treatment of positive operators) to show that the eigenvalue 1 of R_0, which is associated with $g_1(\omega) = 1$, is simple and that all the other eigenvalues satisfy $|\lambda| < 1$. However, we give a proof here that does not use the Krein–Rutman theorem.

If $\{\varphi(x - k)\}_{k \in \mathbb{Z}}$ is a Riesz basis, then, as we have seen above, p_0 is a $\mathcal{C}^\infty$-function that satisfies (4.112) and (4.113). From (4.111) we know that p_0 and m_0 have exactly the same zeros, and hence p_0 has only a finite number of zeros in $[-\pi, \pi]$. This means that p_0 satisfies the hypotheses of Theorem 2.2. On the other hand, we know that the infinite product (4.110) generates a localized multiresolution analysis because $\sum_{l \in \mathbb{Z}} |\hat{\phi}(\omega + 2l\pi)|^2 = 1$. From this and Theorem 2.2 we conclude that there exists no non-trivial cycle $\{\omega_1, \ldots, \omega_n\}$ for the transformation $\omega \mapsto 2\omega$ modulo 2π such that $|p_0(\omega_k)| = 1$. The central argument of the proof is indirect: It proceeds by showing that the assumption that $|\lambda| = 1$ implies the existence of a non-trivial cycle on which $|p_0(\omega_k)| = 1$, and this contradicts Theorem 2.2.

Thus assume that $|\lambda| = 1$ and let g_λ be the associated eigenfunction of R_0. Since $f_\lambda \in F_N$, we have

$$g_\lambda(0) = \frac{f_\lambda(0)}{m(0)} = 0, \tag{4.116}$$

which means that $|g_\lambda|$ is not a constant. (If it were, f_λ would be identically zero and not a true eigenfunction.) Since g_λ is the ratio of two trigonometric polynomials and is not constant, $|g_\lambda|$ attains its maximum in $[0, 2\pi]$ at only a finite number of points. Indeed, the equation

$$g_\lambda(\omega) = C \tag{4.117}$$

implies that

$$|f_\lambda(\omega)|^2 - C^2|m(\omega)|^2 = 0, \tag{4.118}$$

which is satisfied for either all $\omega \in [0, 2\pi]$ or for only a finite number of ω since $|f_\lambda|^2 - C^2|m|^2$ is a trigonometric polynomial.

Choose ω_0 to be a point in $[0, 2\pi]$ such that $|g_\lambda(\omega_0)| = \max g_\lambda(\omega)$. Then the assumption that $|\lambda| = 1$, combined with the equation $|p_0(\omega)|^2 + |p_0(\omega + \pi)|^2 = 1$, implies that

$$|g_\lambda(\omega_0)| = |\lambda g_\lambda(\omega_0)|$$
$$\leq \left|p_0\left(\frac{\omega_0}{2}\right)\right|^2\left|g_\lambda\left(\frac{\omega_0}{2}\right)\right| + \left|p_0\left(\frac{\omega_0}{2}+\pi\right)\right|^2\left|g_\lambda\left(\frac{\omega_0}{2}+\pi\right)\right|$$
$$\leq \max\left(\left|g_\lambda\left(\frac{\omega_0}{2}\right)\right|, \left|g_\lambda\left(\frac{\omega_0}{2}+\pi\right)\right|\right).$$

This means that at least one of the numbers $\{\frac{\omega_0}{2}, \frac{\omega_0}{2}+\pi\}$ corresponds to a maximum of $|g_\lambda|$ and that the inequalities in the last array are in fact equalities.

We now use the notation and arguments that we used in Chapter 2, Sections 2.3 and 2.4. Thus we let

$$S_0(\omega) = \frac{\omega}{2} \quad \text{and} \quad S_1(\omega) = \frac{\omega}{2} + \pi \qquad (4.119)$$

denote the two possible antecedents in $[0, 2\pi[$ of ω under the transformation $\omega \mapsto 2\omega$ modulo 2π. If we represent the binary expansion of $\frac{\omega}{2\pi}$ for ω in $[0, 2\pi[$ by

$$\frac{\omega}{2\pi} = \sum_{j=1}^{+\infty} \alpha_j 2^{-j}, \quad \alpha_j \in \{0, 1\}, \qquad (4.120)$$

then the action of S_0 adds a 0 to the beginning of the sequence $\{\alpha_j\}_{j\geq 1}$ and the action of S_1 adds a 1.

Since at least one of the numbers $S_0(\omega_0)$ and $S_1(\omega_0)$ corresponds to a maximum of $|g_\lambda|$, we can iterate this process and construct a sequence

$$\omega_j = S_{\epsilon_j}(\omega_{j-1}), \quad \epsilon_j \in \{0, 1\}, \qquad (4.121)$$

such that $|g_\lambda(\omega_j)| = \max|g_\lambda(\omega)|$. This sequence is finite because there are only a finite number of maxima in $[0, 2\pi]$. It is also nontrivial because $g_\lambda(0) = 0$. As in the proof of Lemma 2.6, we conclude that the binary expansion of $\frac{\omega_0}{2\pi}$ must be periodic and that the sequence $\{\omega_j\}_{j\geq 1}$ must be periodic. This means that, for each j, only one of the numbers $S_0(\omega_j)$ and $S_1(\omega_j)$ corresponds to a maximum of $|g_\lambda|$. Said another way, the sequence $\{\epsilon_j\}_{j\geq 1}$ is uniquely determined by the periodic structure of the binary expansion of $\frac{\omega_0}{2\pi}$. Thus the other antecedent of ω_j under the transformation $\omega \mapsto 2\omega$ modulo 2π, namely, $\omega_{j+1} + \pi$ (modulo 2π), cannot be a point where $|g_\lambda|$ assumes its maximum.

By construction, we know that

$$|g_\lambda(\omega_{j+1})| = |p_0(\omega_j)|^2|g_\lambda(\omega_j)| + |p_0(\omega_j+\pi)|^2|g_\lambda(\omega_j+\pi)|. \qquad (4.122)$$

Since $|g_\lambda(\omega_j + \pi)| < |g_\lambda(\omega_j)|$ and $|p_0(\omega_j)|^2 + |p_0(\omega_j + \pi)|^2 = 1$, it follows that

$$|p_0(\omega_j)| = 1 \text{ and } p_0(\omega_j + \pi) = 0 \qquad (4.123)$$

for all $j \geq 0$.

We have shown that the assumption that $|\lambda| = 1$ leads to the existence of a non-trivial cycle $\{\omega_1, \ldots, \omega_n\}$ for the transformation $\omega \mapsto 2\omega$ modulo 2π on which $|p_0(\omega_k)| = 1$. This, however, contradicts ii) of Theorem 2.2, and we conclude that $|\lambda| < 1$. This completes the proof of Proposition 4.9 $\square$

The next proposition forms an essential step in the proof of the main theorem. We state and prove it separately because of its intrinsic interest. It, in turn, is supported by a lemma that is also of independent interest.

We introduce the following notation, which we will use in the remainder of this section. Write

$$m_0(\omega) = \sum_{k=N_1}^{N_2} h_k \, e^{-ik\omega} \quad \text{and} \quad \tilde{m}_0(\omega) = \sum_{l=\tilde{N}_1}^{\tilde{N}_2} \tilde{h}_l \, e^{-il\omega},$$

where we specifically assume that h_{N_1}, h_{N_2}, $\tilde{h}_{\tilde{N}_1}$, and $\tilde{h}_{\tilde{N}_2}$ are not zero.

Lemma 4.6 *Assume that the trigonometric polynomials m_0 and $\tilde{m}_0$ satisfy the duality relation*

$$\overline{m_0(\omega)} \, \tilde{m}_0(\omega) + \overline{m_0(\omega + \pi)} \, \tilde{m}_0(\omega + \pi) = 1. \qquad (4.124)$$

Let $\{N_1, \ldots, N_2\}$ denote the set of integers in $[N_1, N_2]$ such that $k \in \{N_1, \ldots, N_2\}$ implies that $h_k \neq 0$. Define $\{\tilde{N}_1, \ldots, \tilde{N}_2\}$ similarly. Then $\{N_1, \ldots, N_2\} \cap \{\tilde{N}_1, \ldots, \tilde{N}_2\}$ is not empty.

Proof. By integrating both sides of (4.125), we see that

$$\frac{1}{2\pi} \int_0^{2\pi} \overline{m_0(\omega)} \, \tilde{m}_0(\omega) \, d\omega = \frac{1}{2\pi} \int_0^{2\pi} \overline{m_0(\omega + \pi)} \, \tilde{m}_0(\omega + \pi) \, d\omega = \frac{1}{2}.$$

If we express m_0 and $\tilde{m}_0$ in terms of their series representations, we conclude that

$$\frac{1}{2\pi} \int_0^{2\pi} \overline{m_0(\omega)} \, \tilde{m}_0(\omega) \, d\omega = \sum_{k,l} \overline{h_k} \, \tilde{h}_l \, \delta_{k,l} = \sum_{n \in \mathbb{Z}} \overline{h_n} \, \tilde{h}_n \qquad (4.125)$$

where the first sum is taken over all $k \in \{N_1, \ldots, N_2\}$ and all $l \in \{\tilde{N}_1, \ldots, \tilde{N}_2\}$. If these two sets do not intersect, then the sums in (4.125) are clearly zero. Since this sum equals 1/2, the two sets must have at least one point in common. □

A corollary of this lemma is that the two intervals $[N_1, N_2]$ and $[\tilde{N}_1, \tilde{N}_2]$ must intersect in at least one point. In particular,

$$N_2 - \tilde{N}_1 \geq 0 \quad \text{and} \quad \tilde{N}_2 - N_1 \geq 0. \tag{4.126}$$

Although it is not apparent at this stage, it will turn out that

$$N_2 - \tilde{N}_1 > 0 \quad \text{and} \quad \tilde{N}_2 - N_1 > 0, \tag{4.127}$$

which is to say that the two intervals must overlap in a set of positive measure.

To avoid trivial cases, which do not lead to L^2 scaling functions, we assume that the filters are at least of length 2, i.e., $N_2 - N_1 \geq 1$ and $\tilde{N}_2 - \tilde{N}_1 \geq 1$. We use this in the following proposition, which is a converse of Proposition 4.5.

Proposition 4.10 *Assume that φ and $\tilde{\varphi}$ are generated by filters m_0 and $\tilde{m}_0$ and that φ and $\tilde{\varphi}$ are in $L^2(\mathbb{R})$. Then the relation*

$$\langle \psi(x-k), \tilde{\psi}(x-k') \rangle = \delta_{k,k'} \tag{4.128}$$

on the wavelets ψ and $\tilde{\psi}$ implies that the same condition holds for the scaling functions φ and $\tilde{\varphi}$, namely, that

$$\langle \varphi(x-k), \tilde{\varphi}(x-k') \rangle = \delta_{k,k'}. \tag{4.129}$$

Proof. We introduce the notation

$$\alpha(\omega) = \sum_{l \in \mathbb{Z}} \overline{\hat{\varphi}(\omega + 2l\pi)} \, \widehat{\tilde{\varphi}}(\omega + 2l\pi).$$

The assumption that φ and $\tilde{\varphi}$ are in $L^2(\mathbb{R})$ and the fact that they have compact support guarantees that the sum defining α is in $L^\infty(\mathbb{R})$ and is equal (almost everywhere) to a trigonometric polynomial (Lemma 4.1). The same is true for the corresponding sum involving ψ and $\tilde{\psi}$. In what follows, we identify these sums with the trigonometric polynomials and drop the 'almost everywhere' expression. Thus, proving (4.129) is equivalent to showing that $\alpha(\omega) = 1$ for all ω.

With this convention, (4.128) is equivalent to saying that

$$\sum_{l \in \mathbb{Z}} \overline{\hat{\psi}(\omega + 2l\pi)} \, \widehat{\tilde{\psi}}(\omega + 2l\pi) = 1$$

for all ω. Since this is true for all ω, we can replace ω by 2ω and

write
$$\sum_{l \in \mathbb{Z}} \overline{\hat{\psi}(2\omega + 2l\pi)} \, \widehat{\tilde{\psi}}(2\omega + 2l\pi) = 1, \qquad (4.130)$$
which becomes
$$\overline{m_0(\omega)}\,\tilde{m}_0(\omega)\,\alpha(\omega+\pi) + \overline{m_0(\omega+\pi)}\,\tilde{m}_0(\omega+\pi)\,\alpha(\omega) = 1 \quad (4.131)$$
by writing $\hat{\psi}$ in terms of $\hat{\varphi}$ and $\widehat{\tilde{\psi}}$ in terms of $\widehat{\tilde{\varphi}}$ using (4.17) and (4.18).

This is how the rest of the proof will go: We multiply m_0, $\tilde{m}_0$, and α by suitable powers of $e^{i\omega}$ to transform the equations (4.124) and (4.131) into polynomial equations in $e^{i\omega} = z$. We then extend these relations, which hold for $|z| = 1$, to the complex plane and use information about the degrees of the polynomials to show that (4.131) is possible only if $\alpha(\omega) = 1$ for all ω.

We begin with α. From Lemma 4.1, α is a trigonometric polynomial $\alpha(\omega) = \sum \alpha_k e^{-ik\omega}$ where $\alpha_k = \int \tilde{\varphi}(x)\overline{\varphi(x-k)}\,dx$. Furthermore, we know that the support of φ is in $[N_1, N_2]$ and that the support of $\tilde{\varphi}$ is in $[\tilde{N}_1, \tilde{N}_2]$. From this we conclude that $\alpha_k = 0$ for $k \geq \tilde{N}_2 - N_1$ and for $k \leq \tilde{N}_1 - N_2$. Thus we can represent α as

$$\alpha(\omega) = \sum_{k=\tilde{N}_1-N_2+1}^{\tilde{N}_2-N_1-1} \alpha_k \, e^{-ik\omega}.$$

To ease notation let $M = \tilde{N}_2 - N_1$ and $N = N_2 - N_1 + \tilde{N}_2 - \tilde{N}_1$. Then
$$e^{i(M-1)\omega}\,\alpha(\omega) = A(e^{i\omega})$$
is a polynomial in $e^{i\omega}$ and the degree of $A \leq N - 2$.

In the same way, we see that
$$e^{-iN_1\omega}\,\overline{m_0(\omega)} = \overline{h}_{N_1} + \cdots + \overline{h}_{N_2} e^{i(N_2-N_1)\omega}$$
is a polynomial in $e^{i\omega}$ of degree $N_2 - N_1$ and
$$e^{i\tilde{N}_2\omega}\,\tilde{m}_0(\omega) = \tilde{h}_{\tilde{N}_1} e^{i(\tilde{N}_2-\tilde{N}_1)\omega} + \cdots + \tilde{h}_{\tilde{N}_2}$$
is a polynomial in $e^{i\omega}$ of degree $\tilde{N}_2 - \tilde{N}_1$. Thus,
$$e^{iM\omega}\,\overline{m_0(\omega)}\,\tilde{m}_0(\omega) = \overline{h}_{N_1}\tilde{h}_{\tilde{N}_2} + \cdots + \overline{h}_{N_2}\tilde{h}_{\tilde{N}_1} e^{iN\omega}$$
is a polynomial in $e^{i\omega}$ of degree N. We denote this polynomial by Γ and write $z = e^{i\omega}$. With this notation we have
$$\overline{m_0(\omega)}\tilde{m}_0(\omega) = z^{-M}\,\Gamma(z)$$

and
$$\overline{m_0(\omega+\pi)}\tilde{m}_0(\omega+\pi) = (-z)^{-M}\Gamma(-z).$$
Furthermore, by substitution, (4.124) becomes
$$z^{-M}\Gamma(z) + (-z)^{-M}\Gamma(-z) = 1 \qquad (4.132)$$
and (4.131) becomes
$$z^{-M}\Gamma(z)(-z)^{-M+1}A(-z) + (-z)^{-M}\Gamma(-z)z^{-M+1}A(z) = 1. \qquad (4.133)$$
By subtracting the latter from the former and multiplying by $z^M(-z)^M$ we conclude that
$$\Gamma(z)[(-z)^M + zA(-z)] + \Gamma(-z)[z^M - zA(z)] = 0. \qquad (4.134)$$

This relation is a polynomial equation that holds for all $|z|=1$. Hence, we can, and do, extend (4.134) to the complex plane. If we multiply (4.132) by z^M it becomes
$$\Gamma(z) + (-1)^M \Gamma(-z) = z^M, \qquad (4.135)$$
and it also holds for all $z \in \mathbb{C}$.

This last equation tells us that a zero of $\Gamma(z)$ cannot be a zero of $\Gamma(-z)$: If $\Gamma(z_0) = 0$, then $(-1)^M \Gamma(z_0) = (z_0)^M$. But $z_0 \neq 0$ since $\Gamma(0) = \overline{h}_{N_1} \tilde{h}_{\tilde{N}_2} \neq 0$.

The argument proceeds as follows: The zeros of Γ, counted with their multiplicity, are zeros of $z^M - zA(z) = B(z)$. Thus, either $B(z) \equiv 0$ or the degree of B $\geq$ the degree of Γ. We know that the degree of Γ is exactly N and that the degree of A is $\leq N-2$. Unfortunately, at this point, we cannot rule out the possibility that $M = N$. In particular, $N = M + N_2 - \tilde{N}_1$ and we do not know a priori that $N_2 - \tilde{N}_1 > 0$. Thus, we cannot conclude that $B(z) \equiv 0$ because its degree is less than the degree of Γ. We take care of this problem with a small detour.

Assume that $N_2 - \tilde{N}_1 = 0$. Because we have also assumed that both $N_2 - N_1$ and $\tilde{N}_2 - \tilde{N}_1$ are ≥ 1, we must have $M \geq 2$. Thus we can divide the left-hand side of (4.134) by z and still retain a polynomial relation. Now we know that the degree of $B(z)/z$ is less than N and conclude from the above argument that $B(z)/z \equiv 0$. This means that $z^{M-1} = A(z)$ for all z, which means that $\alpha(\omega) = 1$ for all ω. But this implies that $\alpha_0 = \int \tilde{\varphi}(x)\overline{\varphi(x)}\,dx = 1$, from which we conclude that $[N_1, N_2]$ and $[\tilde{N}_1, \tilde{N}_2]$ must intersect in more than one point. Thus both $N_2 - \tilde{N}_1$ and $\tilde{N}_2 - N_1$ are greater than zero.

The result of this detour is that M is indeed $< N$ and that the degree of B is less than the degree of Γ. Returning to the central argument, we conclude that $B(z) \equiv 0$ and that $\alpha(\omega) = 1$ for all ω. This proves the proposition and prepares the way for an easy proof of the main theorem. □

Theorem 4.2 *Let m_0 and $\tilde{m}_0$ be trigonometric polynomials that satisfy the duality relation*

$$\overline{m_0(\omega)}\,\tilde{m}_0(\omega) + \overline{m_0(\omega+\pi)}\,\tilde{m}_0(\omega+\pi) \;=\; 1\,.$$

and the 'low-pass' conditions

$$m_0(0) = \tilde{m}_0(0) = 1 \quad \text{and} \quad m_0(\pi) = \tilde{m}_0(\pi) = 0\,.$$

Let P_0 and $\tilde{P}_0$ be the transition operators associated with m_0 and $\tilde{m}_0$, respectively, and denote by $\rho(P_0)$ and $\rho(\tilde{P}_0)$ the spectral radii of P_0 and $\tilde{P}_0$ restricted to F_N and to $\tilde{F}_N$. Then the dual filters m_0 and $\tilde{m}_0$ generate unconditional biorthogonal bases of compactly supported wavelets for $L^2(\mathbb{R})$ if and only if

$$|\rho(P_0)| < 1 \quad \text{and} \quad |\tilde{\rho}(P_0)| < 1\,.$$

Proof. We first prove the necessity of these conditions. Thus assume that m_0 and $\tilde{m}_0$ lead to unconditional biorthogonal wavelet bases $\{\psi_{j,k}\}_{j,k\in\mathbb{Z}}$ and $\{\tilde{\psi}_{j,k}\}_{j,k\in\mathbb{Z}}$. Then

$$\langle \psi(x-k), \tilde{\psi}(x-k')\rangle \;=\; \delta_{k,k'}\,,$$

which by Proposition 4.10 implies that

$$\langle \varphi(x-k), \tilde{\varphi}(x-k')\rangle \;=\; \delta_{k,k'}\,,$$

or equivalently that

$$1 \;=\; \sum_{l\in\mathbb{Z}} \overline{\hat{\varphi}(\omega+2l\pi)}\,\hat{\tilde{\varphi}}(\omega+2l\pi)\,.$$

An application of the Cauchy–Schwarz inequality shows that

$$1 \;\leq\; \Big(\sum_{l\in\mathbb{Z}} |\hat{\varphi}(\omega+2l\pi)|^2\Big)\Big(\sum_{l\in\mathbb{Z}} |\hat{\tilde{\varphi}}(\omega+2l\pi)|^2\Big)\,.$$

This means that the trigonometric polynomials $\sum_{l\in\mathbb{Z}} |\hat{\varphi}(\omega+2l\pi)|^2$ and $\sum_{l\in\mathbb{Z}} |\hat{\tilde{\varphi}}(\omega+2l\pi)|^2$ are bounded away from zero. Thus there are four strictly positive constants C_1, C_2, $\tilde{C}_1$, and $\tilde{C}_2$ such that

$$C_1 \;\leq\; \sum_{l\in\mathbb{Z}} |\hat{\varphi}(\omega+2l\pi)|^2 \;\leq\; C_2\,,$$

and
$$\tilde{C}_1 \leq \sum_{l \in \mathbb{Z}} |\widehat{\tilde{\varphi}}(\omega + 2l\pi)|^2 \leq \tilde{C}_2.$$

These two relations are equivalent to saying that $\{\varphi(x-k)\}_{k \in \mathbb{Z}}$ and $\{\tilde{\varphi}(x-k)\}_{k \in \mathbb{Z}}$ are Riesz bases for the spaces they generate. From Proposition 4.9 we conclude that

$$|\rho(P_0)| < 1 \quad \text{and} \quad |\tilde{\rho}(P_0)| < 1,$$

which proves the theorem in one direction.

To prove the result in the other direction, assume that the spectral radii are both less than 1. Then from Proposition 4.8 we know that $\hat{\varphi}$ and $\widehat{\tilde{\varphi}}$ are in $L^2(\mathbb{R})$ and that the truncated products $\hat{h}_n$ and $\widehat{\tilde{h}}_n$ tend to $\hat{\varphi}$ and $\widehat{\tilde{\varphi}}$ in $L^2(\mathbb{R})$. Knowing this, we recover the conclusions of Propositions 4.2 through 4.6. The assumption on the spectral radii also implies the conclusions of Lemma 4.4, which means that the relations (4.80) and (4.81) hold. But this is exactly what is needed to prove Proposition 4.7 and thus to recover the conclusions of Theorem 4.1, namely, that the families $\{\psi_{j,k}\}_{j,k \in \mathbb{Z}}$ and $\{\tilde{\psi}_{j,k}\}_{j,k \in \mathbb{Z}}$ are unconditional biorthogonal bases for $L^2(\mathbb{R})$. This proves the theorem in the other direction and completes the proof. □

This proof completes our theoretical development of biorthogonal wavelet bases. It is now time to present some examples of applications of these results to show the considerable flexibility that biorthogonal wavelets offer for numerical applications. The task is to look for discrete, finite impulse response filters that have interesting properties and that satisfy relation (4.14) and the hypotheses of Proposition 4.1.

4.5 Examples and applications

We present two important classes of filters: the filters introduced by P. J. Burt and E. H. Adelson and spline filters.

4.5.1 Burt and Adelson's filters

In 1983, P. J. Burt and E. H. Adelson presented a multiresolution analysis technique that was developed essentially for digital image processing. This is the process, which differs from subband coding:

- One applies a low-pass filter to the original image and then

EXAMPLES AND APPLICATIONS 145

decimates the result (in the ratio of 1 pixel out of 4 in the case of images).

• Next, one interpolates with the same filter and then encodes the difference between the original signal and the approximation to the signal that one has obtained.

One quickly sees that, if the initial image consists of N pixels, then the multiscale representation will have approximately N' pixels where $N' = N + \frac{N}{4} + \frac{N}{16} + \cdots = \frac{4}{3}N$. This factor shows that one has kept too much information. The problem is that, while the first approximation takes only $\frac{N}{4}$ pixels, coding the 'details' still takes N pixels. Thus, after the first pass $N + \frac{N}{4}$ pixels are needed. At the next step the count is $N + \frac{N}{4} + \frac{N}{16}$, and so on. With an efficient scheme, coding the details at the first stage should take only $\frac{3}{4}N$ pixels.

Subband coding avoids this difficulty, but, unfortunately, the filters used by Burt and Adelson cannot be fitted into the context of orthonormal wavelets. This is because the Burt and Adelson filters must be real, finite, and symmetric to be well adapted to image processing, and we know that the only CQF with these properties is the one associated with the Haar system. We also want the associated scaling function φ to have some minimum amount of regularity, for, as we saw in Chapter 3, this is related to the algorithm and the quality of approximation. We will see concrete examples of this in Chapter 5. Clearly, if we want our filters to have all of these qualities, it is necessary to abandon orthonormal wavelets.

We will see that Burt and Adelson's filters fit perfectly into the biorthogonal setting. This idea is due to Michel Barlaud and has been developed in collaboration with Ingrid Daubechies for image compression.

The construction of these filters follows naturally from the constraints cited above. The function m_0 must satisfy several properties:

• $m_0(0) = 1$ and $m_0(\pi) = 0$, where the first equation means that the filtering is an 'averaging' process.

• The Fourier coefficients are real, symmetric about the origin, and finite in number.

• So that φ will have a certain regularity, we require that m_0 be divisible by $(\frac{1+e^{i\omega}}{2})$. Because of the symmetry, one can deduce that m_0 is then divisible by $\cos^2(\omega/2) = (\frac{1+e^{i\omega}}{2})(\frac{1+e^{-i\omega}}{2})$.

- Finally, one would like to have a choice among a number of filters.

It is clear that the simplest example, the one with three coefficients, is given by $m_0(\omega) = \cos^2(\omega/2)$.

The filters with five coefficients are more interesting. In this case we have a one-parameter family given by

$$m_{0,a}(\omega) = \cos^2\left(\frac{\omega}{2}\right)[4a - 1 + (2 - 4a)\cos\omega], \qquad (4.136)$$

which we write as

$$m_{0,a}(\omega) = \cos^2\left(\frac{\omega}{2}\right) f_a(\omega). \qquad (4.137)$$

This parametric representation corresponds to the following expressions for the coefficients of $m_{0,a}$:

$$h_0 = a; \quad h_{-1} = h_1 = \frac{1}{4}; \quad h_{-2} = h_2 = \frac{1}{4} - \frac{a}{2}. \qquad (4.138)$$

This is the family that was chosen by Burt and Adelson, and we will examine several cases in detail. As a first step, we wish to construct the function $\tilde{m}_{0,a}$, which, with $m_{0,a}$, will satisfy property (4.14). Later, we will investigate the regularity of the associated scaling functions φ_a and $\tilde{\varphi}_a$.

Again, we want $\tilde{m}_{0,a}(\omega) = \sum_{l=-N}^{N} \tilde{h}_l e^{-il\omega}$ to be real, symmetric, and divisible by $\cos^2(\omega/2)$. This filter has $2N + 1$ real coefficients that are symmetric about the origin, which means that we must determine $N + 1$ unknown coefficients. Equation (4.14) implies immediately that $\tilde{m}_{0,a}(0) = 1$. Thus this relation is not independent of (4.14), which is expressed in terms of the coefficients as

$$2 \sum_{k=-2}^{2} h_k \tilde{h}_{k+2n} = \delta_{0,n} \qquad \text{for all} \qquad n \in \mathbb{Z}. \qquad (4.139)$$

This relation implies that N must be odd, for if it is not, taking $n = \frac{N}{2} - 1$ implies that $h_2 \tilde{h}_N = 0$. This means that $\tilde{h}_N = 0$ and contradicts the definition of N. Because of symmetry, this form of (4.14) also yields a system of $\frac{N+3}{2}$, a priori, independent linear equations. We add to these the constraint $\tilde{m}_{0,a}(\pi) = 0$, which ensures that $\tilde{m}_{0,a}$ is divisible by $\cos^2(\omega/2)$. Thus we have $\frac{N+3}{2} + 1$ equations in $N + 1$ unknowns. Clearly the number of unknowns grows faster than the number of equations, with the crossover being at $N = 3$. The choice $N = 3$ seems reasonable if we wish to solve this system and find dual filters $\tilde{m}_{0,a}$ having the least number

EXAMPLES AND APPLICATIONS

of coefficients. With this choice, the system of equations can be written as

$$\begin{bmatrix} 2a & 1 & 1-2a & 0 \\ 0 & 0 & 1-2a & 1 \\ 1-2a & 1 & 4a & 1 \\ 1 & -2 & 2 & -2 \end{bmatrix} \begin{bmatrix} \tilde{h}_0 \\ \tilde{h}_1 \\ \tilde{h}_2 \\ \tilde{h}_3 \end{bmatrix} = \begin{bmatrix} 1 \\ 0 \\ 0 \\ 0 \end{bmatrix}. \quad (4.140)$$

A unique solution exists if $a \neq \frac{1}{4}$, and in this case we have

$$\tilde{h}_0 = \frac{1+4a}{4(4a-1)}, \quad \tilde{h}_1 = \frac{-8a^2+18a-5}{8(4a-1)},$$

$$\tilde{h}_2 = \frac{4a-3}{8(4a-1)}, \quad \tilde{h}_3 = \frac{(1-2a)(3-4a)}{8(4a-1)}.$$

The function $\tilde{m}_{0,a}$ can then be written as

$$\tilde{m}_{0,a}(\omega) = \cos^2\left(\frac{\omega}{2}\right) \tilde{f}_a(\omega) \quad (4.141)$$

with

$$\tilde{f}_a(\omega) = \frac{(8a^2-10a+5)-(3-4a)^2 \cos\omega + (3-4a)(1-2a) \cos 2\omega}{4a-1}. \quad (4.142)$$

It is not surprising that this function cannot be defined for the value $a = \frac{1}{4}$. Indeed, in this case we would have $m_{0,\frac{1}{4}}(\omega) = \cos^2\left(\frac{\omega}{2}\right) \cos\omega$, and hence $m_{0,\frac{1}{4}}(\pi/2) = m_{0,\frac{1}{4}}(-\pi/2) = 0$. But this means that both $m_{0,\frac{1}{4}}(\pi/2)$ and $m_{0,\frac{1}{4}}(\pi/2 + \pi)$ equal zero, which is incompatible with (4.14).

This is the appropriate place to discuss the regularity of the scaling functions φ_a and $\tilde{\varphi}_a$ that are defined in terms of the filters $m_{0,a}$ and $\tilde{m}_{0,a}$. Our approach is to use the critical exponents of these filters and the results of Chapter 3 to ensure, in the first place, that we indeed have unconditional biorthogonal bases and then to study the regularity of the scaling functions and the consequent behavior of the subdivision algorithms. Thus, a minimal requirement using this approach is that the critical exponents satisfy the hypotheses of Proposition 4.1. Once these are satisfied, we can try to go further and study the regularity of φ_a and $\tilde{\varphi}_a$.

If we denote the critical exponents of $m_{0,a}$ and $\tilde{m}_{0,a}$ by b_a and $\tilde{b}_a$ then, since $N = 2$, we want to have simultaneously

$$b_a < \frac{3}{2} \quad \text{and} \quad \tilde{b}_a < \frac{3}{2}. \quad (4.143)$$

To investigate the regularity, we use the relation

$$N - b - 1 \leq s_1 \leq \mu, \qquad (4.144)$$

which relates the critical exponent b to the global Hölder exponent μ. (Note that this relation, unlike the consequence of Theorem 3.1, does not use the assumption (3.49), which is generally not satisfied for all a.) Thus, if we wish to conclude something about the regularity of $\tilde{\varphi}_a$ using this technique, we need to have $\tilde{b}_a < 1$.

The investigation of these properties as a function of a turns out to be computationally intensive and rather difficult because the methods for estimating these critical exponents differ according to the value of a.

It appears that the only realistic values for this parameter are between 0.3 and 1. Outside this interval, one observes that the decay of $\hat{\varphi}_a$, or of $\widehat{\tilde{\varphi}}_a$, is poor and that the subdivision algorithm gives very spurious results.

We are going to examine several significant cases, each of which will be dealt with differently depending on the value of a.

Case 1) $a = 3/8 = 0.375$

This case is interesting because it is the one that gives the maximum regularity that can be obtained for the function φ_a. (This can be shown using the techniques presented in Daubechies and Lagarias (1991, 1992).)

Here we have

$$m_{0,\frac{3}{8}}(\omega) = \cos^4\left(\frac{\omega}{2}\right), \qquad (4.145)$$

and the function $\varphi_{\frac{3}{8}}$ is the quadruple convolution product of the step function $\mathbf{1}_{[-\frac{1}{2},\frac{1}{2}]}$ with itself, that is,

$$\varphi_{\frac{3}{8}}(x) = (*)^4 \mathbf{1}_{[-\frac{1}{2},\frac{1}{2}]}(x). \qquad (4.146)$$

Thus $\varphi_{\frac{3}{8}}$ is piecewise a cubic polynomial and globally $\mathcal{C}^2$.

REMARKS

Unfortunately, $\tilde{m}_{0,\frac{3}{8}}$ fails the hypothesis of Proposition 4.1. To see this, we first compute

$$\tilde{f}_{\frac{3}{8}}(\omega) = 4.75 - 4.5 \cos\omega + 0.75 \cos 2\omega. \qquad (4.147)$$

From the result (3.87) we know that

$$\tilde{b}_a \geq \frac{1}{\log 2} \log\left(|\tilde{f}_a(\frac{2\pi}{3})|\right), \qquad (4.148)$$

and another computation shows that $\tilde{f}_{\frac{3}{8}}(\frac{2\pi}{3}) = 6.625$, which is

EXAMPLES AND APPLICATIONS   149

greater than $\sqrt{8} = 2^{3/2}$. Thus $\tilde{b}_{\frac{3}{8}} > \frac{3}{2}$ and (4.143) is not satisfied. We will return to this example in the second part of this section; there we will use a larger number of coefficients to construct a biorthogonal basis. The results of computing $\varphi_{\frac{3}{8}}$ and $\tilde{\varphi}_{\frac{3}{8}}$ with the subdivision algorithm are presented in Figure 4.1. The bad behavior due to the poor decay of $\widehat{\tilde{\varphi}}_{\frac{3}{8}}$ is clearly apparent.

Case 2) $a = \frac{1}{2} = 0.5$

Again the function $m_{0,\frac{1}{2}}$ has a simple expression since $f_{\frac{1}{2}} = 1$. The scaling function $\varphi_{\frac{1}{2}}$ is thus the 'triangle' centered at the origin, which is given by

$$\varphi_{\frac{1}{2}}(x) = (*)^2 \mathbf{1}_{[-\frac{1}{2},\frac{1}{2}]}(x). \tag{4.149}$$

The coefficients of $\tilde{m}_{0,\frac{1}{2}}$ are

$$\tilde{h}_0 = \frac{3}{4}; \quad \tilde{h}_1 = \frac{1}{4}; \quad \tilde{h}_2 = -\frac{1}{8}; \quad \tilde{h}_4 = 0. \tag{4.150}$$

This shows immediately that $\tilde{m}_{0,\frac{1}{2}} = m_{0,\frac{3}{4}}$ and $\tilde{m}_{0,\frac{3}{4}} = m_{0,\frac{1}{2}}$, and that, consequently, the cases $a = \frac{1}{2}$ and $a = \frac{3}{4}$ are equivalent.

We now estimate $\tilde{b}_{\frac{1}{2}}$. Since $\tilde{f}_{\frac{1}{2}}(\omega) = 2 - \cos\omega$, it is clear that $\sup_{\omega \in \mathbb{R}} |\tilde{f}_{\frac{1}{2}}(\omega)| = 3 > \sqrt{8}$. Thus our first estimate of $\tilde{b}_{\frac{1}{2}}$ by $\tilde{b}_{\frac{1}{2},1}$ is inconclusive. To estimate $\tilde{b}_{\frac{1}{2},2} = \frac{1}{2\log 2} \sup_{\omega \in \mathbb{R}} |\tilde{f}_{\frac{1}{2}}(\omega)\tilde{f}_{\frac{1}{2}}(2\omega)|$, we express the product $\tilde{f}_{\frac{1}{2}}(\omega)\tilde{f}_{\frac{1}{2}}(2\omega)$ as a function of the variable $c = \cos\omega$, and study the polynomial

$$\tilde{f}_{\frac{1}{2}}(\omega)\tilde{f}_{\frac{1}{2}}(2\omega) = (2-c)(3-2c^2) = 2c^3 - 4c^2 - 3c + 6$$

between -1 and 1. The maximum occurs at $c_m = \frac{4-\sqrt{34}}{6}$, and one finds that $\sup_{\omega \in \mathbb{R}} |\tilde{f}_{\frac{1}{2}}(\omega)\tilde{f}_{\frac{1}{2}}(2\omega)| \approx 6.486 < 8$. Consequently, $\tilde{b}_{\frac{1}{2}} \leq \tilde{b}_{\frac{1}{2},2} \approx 1.36 < 3/2$, and we recover the results of Proposition 4.1 and the existence of biorthogonal wavelet bases. On the other hand, from (4.148) we see that $\tilde{b}_{\frac{1}{2}} \geq \frac{\log(2.5)}{\log 2} \approx 1.322 > 1$, and we cannot hope to deduce the continuity (or any Hölder regularity) of $\tilde{\varphi}_{\frac{1}{2}}$ using this method.

Case 3) $a = 0.6$

The results in this case are rather surprising, and they are interesting for two reasons:

- First, a computation shows that both $|f_{0.6}(\omega)|$ and $|\tilde{f}_{0.6}(\omega)|$ are < 2 for all ω. This means that $\varphi_{0.6}$ and $\tilde{\varphi}_{0.6}$ are continuous, since both $b_{0.6}$ and $\tilde{b}_{0.6}$ are less than 1.
- Second, the graphs of $\varphi_{0.6}$ and $\tilde{\varphi}_{0.6}$, which are presented on Figure 4.1, are astonishingly similar. An explanation for this is to notice that the coefficients of the filters $m_{0,0.6}$ and $\tilde{m}_{0,0.6}$, which are given in the summary table, are numerically very close. Ingrid Daubechies also remarked that these coefficients are close to those of the CQFs that are used to construct 'coiflets' (see Daubechies (1992)). Coiflets, which are named for Ronald R. Coifman, are a particular case of wavelets whose scaling function φ has, itself, zero moments and exhibits a little more symmetry than the functions constructed in Daubechies (1988).

All of this means that we are dealing with a case that is very close to being orthonormal and that the Riesz constants A and $\tilde{A}$ are, without a doubt, very close to 1. This is in spite of estimates of the type (4.82), which can be fairly poor.

Case 4) $a = 0.7$

In this case we have $\sup_{\omega \in \mathbb{R}} |\tilde{f}_{0.7}(\omega)| < 2$. This ensures the continuity of $\tilde{\varphi}_{0.7}$, which seems to be more regular than $\varphi_{0.7}$ (Figure 4.2).

We see that $f_{0.7}(\omega) = 1.8 - 0.8 \cos \omega$ and, hence, $|f_{0.7}(\omega)| < \sqrt{8}$. This shows that the hypotheses of Proposition 4.1 are satisfied. On the other hand, $f_{0.7}(\frac{2\pi}{3}) = 2.2$. Thus $b_{0.7} > 1$ and we cannot deduce the continuity of $\varphi_{0.7}$ with this method. Better estimates can be obtained by using the results of Daubechies and Lagarias (1991, 1992).

Case 5) $a = 0.8$

This example leads naturally to the conjecture that we proved in Proposition 3.5 for CQFs of type (3.50).

As a first step, it is easy to see that $|\tilde{f}_{0.8}(\omega)| < 2$ and, hence, that the function $\tilde{\varphi}_{0.8}$ presents no problems. However, when we examine $f_{0.8}(\omega) = 2.2 - 1.2 \cos \omega$ we see immediately that its maximum is 3.4, which is clearly greater than $\sqrt{8}$. This leads us to examine the products $f_{0.8,j}(\omega) = \prod_{k=0}^{j} f_{0.8}(2^k \omega)$ with the hope that $b_{0.8,j}$ will be less than $\frac{3}{2}$ for some relatively small j. Note that $f_{0.8}(\frac{2\pi}{3}) = 2.8$, which is less than $\sqrt{8}$ but relatively close to it. When this happens,

EXAMPLES AND APPLICATIONS 151

we know from experience that one can expect a certain amount of difficulty in establishing (4.143).

As before, we write the products $f_{0.8,j}$ in terms of the variable $c = \cos\omega$ and evaluate the maximum of these functions between -1 and 1. These were computed numerically using a program designed for the purpose. One finds that

$$\sup |f_{0.8,1}(\omega)| \approx 8.15 > (\sqrt{8})^2,$$
$$\sup |f_{0.8,3}(\omega)| \approx 65 > (\sqrt{8})^4,$$
$$\sup |f_{0.8,5}(\omega)| \approx 513 > (\sqrt{8})^6,$$

and finally that

$$\sup |f_{0.8,7}(\omega)| \approx 4029 < (\sqrt{8})^8.$$

It has been necessary to go all the way to $c = 7$ to establish (4.143)! By examining the graphs of the $f_{0.8,j}$, one notices that the maxima are attained on a sequence of points that tend toward $j = -\frac{1}{2}$, which is to say, toward $\omega = \frac{2\pi}{3}$. This leads us to conjecture that

$$2^{b_{0.8}} = f_{0.8}\left(\frac{2\pi}{3}\right) = 2.8 \; (< \sqrt{8}). \tag{4.151}$$

We have in any case shown that (4.143) holds. The graphs of $\varphi_{0.8}$ and $\tilde{\varphi}_{0.8}$ are shown in Figure 4.2.

Case 6) $a = 1$

In this case $|\tilde{f}_1(\omega)| < 2$. Unfortunately, $f_1\left(\frac{2\pi}{3}\right) = 5$, and thus there is no hope of obtaining the inequality (4.143) for b_1. The results of the subdivision algorithms are nevertheless shown in Figure 4.2.

Table 4.1 displays a summary of the results for each value of the parameter a that we have examined.

4.5.2 Spline filters

These filters appear in the theory of spline functions with equally spaced knots, i.e., functions in $\mathcal{C}^{N-1}$ that coincide with polynomials of degree N on intervals of the type $[k\tau, (k+1)\tau]$, $k \in \mathbb{Z}$. Briefly, this is the background: The spline functions of degree N with compact support and $\tau = 1$ generate a linear space V_0^N in $L^2(\mathbb{R})$. This leads to a multiresolution analysis $\{V_j^N\}_{j \in \mathbb{Z}}$ by defining $f(x) \in V_j^N$ if and only if $f(2x) \in V_{j+1}^N$. Work by G. Battle and P.-G. Lemarié (see Meyer (1990)) shows that the orthonormal wavelets that one

Table 4.1 *Summary of Burt and Adelson's filters*

Case: $a = \frac{3}{8}$

$h_0 = \frac{3}{8}$ $\quad h_{\pm 1} = \frac{1}{4}$ $\quad h_{\pm 2} = \frac{1}{16}$

$\tilde{h}_0 = \frac{5}{4}$ $\quad \tilde{h}_{\pm 1} = \frac{5}{32}$ $\quad \tilde{h}_{\pm 2} = -\frac{3}{8}$ $\quad \tilde{h}_{\pm 3} = \frac{3}{32}$

φ is a $\mathcal{C}^2$ basic cubic spline, but $\widehat{\tilde{\varphi}}$ does not have the required decay at $\pm\infty$.

Case: $a = \frac{1}{2}$

$h_0 = \frac{1}{2}$ $\quad h_{\pm 1} = \frac{1}{4}$ $\quad h_{\pm 2} = 0$

$\tilde{h}_0 = \frac{3}{4}$ $\quad \tilde{h}_{\pm 1} = \frac{1}{4}$ $\quad \tilde{h}_{\pm 2} = -\frac{1}{8}$ $\quad \tilde{h}_{\pm 3} = 0$

φ is the 'triangle' function, $\widehat{\tilde{\varphi}}$ has the required decay (better than $|\omega|^{-1/2}$), and $m_{0,\frac{1}{2}} = m_{0,\frac{3}{4}}$.

Case: $a = \frac{6}{10}$

$h_0 = \frac{6}{10}$ $\quad h_{\pm 1} = \frac{1}{4}$ $\quad h_{\pm 2} = -\frac{1}{20}$

$\tilde{h}_0 = 0.607$ $\quad \tilde{h}_{\pm 1} = 0.261$ $\quad \tilde{h}_{\pm 2} = -0.054$ $\quad \tilde{h}_{\pm 3} = -0.011$

φ and $\tilde{\varphi}$ are both continuous and their graphs are quite similar. This case is close to being orthonormal.

Case: $a = \frac{7}{10}$

$h_0 = \frac{7}{10}$ $\quad h_{\pm 1} = \frac{1}{4}$ $\quad h_{\pm 2} = -\frac{1}{10}$

$\tilde{h}_0 = 0.528$ $\quad \tilde{h}_{\pm 1} = 0.256$ $\quad \tilde{h}_{\pm 2} = -0.002$ $\quad \tilde{h}_{\pm 3} = 0.006$

$\tilde{\varphi}$ is continuous, and $\hat{\varphi}$ and $\widehat{\tilde{\varphi}}$ have the required decay.

Case: $a = \frac{8}{10}$

$h_0 = \frac{8}{10}$ $\quad h_{\pm 1} = \frac{1}{4}$ $\quad h_{\pm 2} = -\frac{3}{20}$

$\tilde{h}_0 = 0.478$ $\quad \tilde{h}_{\pm 1} = 0.243$ $\quad \tilde{h}_{\pm 2} = 0.011$ $\quad \tilde{h}_{\pm 3} = 0.007$

$\tilde{\varphi}$ is continuous, and $\hat{\varphi}$ has the necessary decay, but just barely.

Case: $a = 1$

$h_0 = 1$ $\quad h_{\pm 1} = \frac{1}{4}$ $\quad h_{\pm 2} = -\frac{1}{4}$

$\tilde{h}_0 = \frac{5}{12}$ $\quad \tilde{h}_{\pm 1} = \frac{5}{24}$ $\quad \tilde{h}_{\pm 2} = \frac{1}{24}$ $\quad \tilde{h}_{\pm 3} = \frac{1}{24}$

$\tilde{\varphi}$ is continuous, but $\hat{\varphi}$ decays slower than $|\omega|^{-1/2}$.

EXAMPLES AND APPLICATIONS

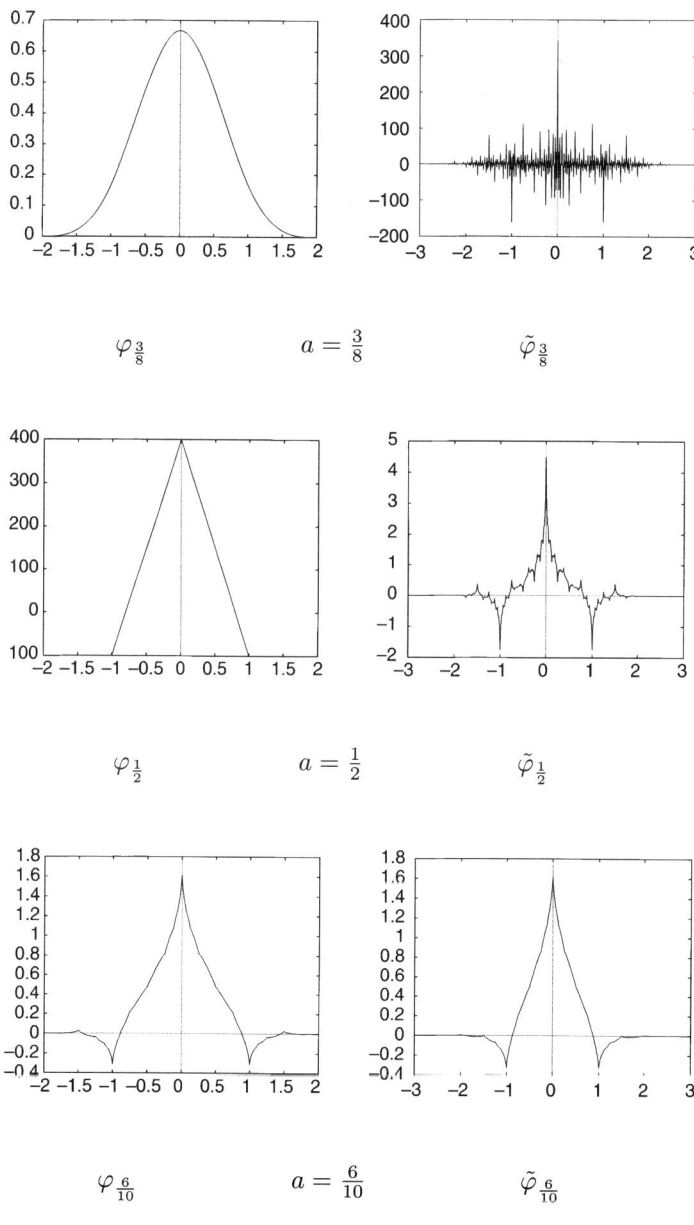

Figure 4.1

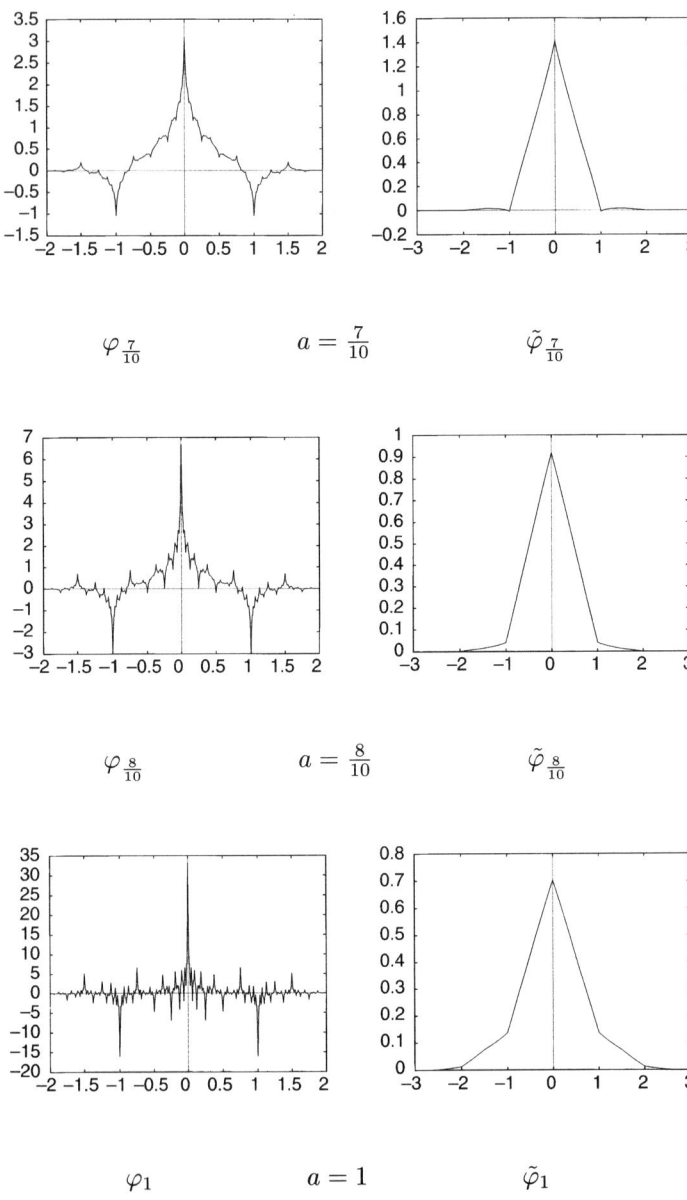

Figure 4.2

EXAMPLES AND APPLICATIONS   155

obtains from this multiresolution analysis, when $N \geq 1$, do not have compact support and, hence, that the corresponding filters are not finite. The wavelets do have exponential decay, however.

Our interest in these multiresolution analyses focuses on the basic spline functions $\phi_N = (*)^{N+1} \mathbf{1}_{[0,1]}$. The ϕ_N and their translates form the simplest Riesz bases for the spaces V_0^N, and, in fact, ϕ_N is usually taken as the starting point for constructing V_0^N. Except for the case $N = 0$, which gives the Haar system, the family $\{\phi_N(x - k)\}_{k \in \mathbb{Z}}$ is not orthonormal. Nevertheless, the functions ϕ_N are of interest to us in the context of biorthogonal wavelet bases, and this is so for several reasons: The ϕ_N are in $\mathcal{C}^{N-1}$, the compact support of ϕ_N is $[0, N + 1]$, and $\phi_N(x) > 0$ for $x \in]0, N + 1[$. Furthermore, ϕ_N is symmetric about $\frac{N+1}{2}$ so the function φ_N defined by $\varphi_N(x) = \phi_N(x + \frac{N+1}{2})$ is symmetric about the origin. When N is odd, $\frac{N+1}{2}$ is an integer translate, and the families $\{\varphi_N(x - k)\}_{k \in \mathbb{Z}}$ and $\{\phi_N(x - k)\}_{k \in \mathbb{Z}}$ are the same Riesz basis for V_0^N. V_0^N is thus generated by the symmetric function φ_N. This is not the case if N is even, where the best we can say is that the generator $\varphi_N(x) = \phi(x + \frac{N}{2})$ is symmetric about $x = \frac{1}{2}$. We now examine these function in the spectral domain.

Since ϕ_N is the $(N + 1)$-fold convolution of $\mathbf{1}_{[0,1]}$, we see that the corresponding filter is given by $\left(\frac{1+e^{-i\omega}}{2}\right)^{N+1}$. As in the time domain, there are two cases:

• The case where N is odd is most often considered in practice. Here we denote the filter associated with φ_N by $m_{0,N}$. Then $\hat{\varphi}_N(\omega) = e^{i\frac{N+1}{2}\omega} \hat{\phi}_N(\omega)$, and

$$\begin{aligned} m_{0,N}(\omega) &= e^{i\frac{N+1}{2}\omega} \left(\frac{1+e^{-i\omega}}{2}\right)^{N+1} \\ &= \left(\frac{1+e^{i\omega}}{2}\right)^{\frac{N+1}{2}} \left(\frac{1+e^{-i\omega}}{2}\right)^{\frac{N+1}{2}} \\ &= \left(\cos\left(\frac{\omega}{2}\right)\right)^{N+1}. \end{aligned}$$

In this case, there exists a unique element in V_0^N, called the 'Lagrangian spline' or 'fundamental spline interpolant,' whose value is 1 at the origin and 0 at the other integers. Furthermore, the operator $f \mapsto \{f(k)\}_{k \in \mathbb{Z}}$ is bijective, and one can interpolate in V_0^N all sequences in $l^2(\mathbb{Z})$. Since our main interest is projection rather than interpolation, we will defer further discussion of these points until the end of the chapter.

- The case where N is even is generally not much considered because the sequences of $l^2(\mathbb{Z})$ cannot be interpolated in V_0^N. Thus a Lagrangian spline does not exist, and, in addition, the function $\varphi_N(x) = \phi_N(x + \frac{N}{2})$ is no longer even but only symmetric about $x = \frac{1}{2}$. The filter to consider is then

$$m_{0,N}(\omega) = e^{-i\frac{\omega}{2}} \left(\cos\left(\frac{\omega}{2}\right)\right)^{N+1}. \quad (4.152)$$

We are interested in the biorthogonal constructions that can be achieved by starting with these functions because they provide multiresolution analyses in the spline function spaces by using finite impulse response filters. This is impossible in the orthonormal case.

To construct the $\tilde{m}_{0,N}$ systematically, we will use the identity (4.153) below that was introduced by Ingrid Daubechies (see Chapter 3) for the construction of orthonormal wavelets with compact support. This is the situation: For each natural number L, there is a unique polynomial of degree $L-1$ that satisfies the equation

$$y^L P_L(1-y) + (1-y)^L P_L(y) = 1, \quad (4.153)$$

and P_L is given by the expression

$$P_L(y) = \sum_{j=0}^{L-1} \binom{L-1+j}{j} y^j. \quad (4.154)$$

(The existence and uniqueness of P_L is a special case of Bezout's theorem (see Daubechies (1992)).) If N is odd and $m_{0,N}(\omega)$ is $\left(\cos\left(\frac{\omega}{2}\right)\right)^{N+1}$, and if we introduce the variable $y = \sin^2\left(\frac{\omega}{2}\right)$, a solution of (4.153) is a solution of (4.14). Thus it would seem that a reasonable choice for $\tilde{m}_{0,N}$ would be $\tilde{m}_{0,N}(\omega) = P_{\frac{N+1}{2}}\left(\sin^2\left(\frac{\omega}{2}\right)\right)$. Unfortunately, this filter is not low-bass, and thus does not satisfy the hypotheses of Proposition 4.1, since $\tilde{m}_{0,N}(\pi) = P_{\frac{N+1}{2}}(1) > 0$.

To introduce zeros at $\omega = \pi$, we will use (4.153) slightly differently; this will also provide a solution in case N is even. Quite simply we take L to be larger than $\frac{N+1}{2}$. We then write

$$\left(\cos\left(\frac{\omega}{2}\right)\right)^{2L} P_L\left(\sin^2\left(\frac{\omega}{2}\right)\right) + \left(\sin\left(\frac{\omega}{2}\right)\right)^{2L} P_L\left(\cos^2\left(\frac{\omega}{2}\right)\right) = 1,$$

which can be decomposed into $m_{0,N}$ and $\tilde{m}_{0,N}$ be writing $2L$ as $(N+1) + (2L - N - 1)$. This leads to the following choices for $\tilde{m}_{0,N}$:

EXAMPLES AND APPLICATIONS   157

- If N is odd
$$\tilde{m}_{0,N}(\omega) = \left(\cos(\frac{\omega}{2})\right)^{2L-N-1} P_L\left(\sin^2(\frac{\omega}{2})\right). \qquad (4.155)$$

- If N is even
$$\tilde{m}_{0,N}(\omega) = e^{-i\frac{\omega}{2}} \left(\cos(\frac{\omega}{2})\right)^{2L-N-1} P_L\left(\sin^2(\frac{\omega}{2})\right). \qquad (4.156)$$

In this construction, the integer L can be chosen to be arbitrarily large. This allows us to satisfy the hypotheses of Proposition 4.1. Specifically, we know from Theorem 3.1 that $s_\infty = 2L - N - 1 - \tilde{b}(L)$ where $\tilde{b}(L)$ is the critical exponent of $\tilde{m}_{0,N}$, which now depends on the parameter L. We also know from Proposition 3.5 (3.90) that

$$\tilde{b}(L) = \frac{2}{\log 2} \log\left(\left|p_L(\frac{2\pi}{3})\right|\right),$$

and from (3.101) that

$$\frac{3^{L-1}}{\sqrt{L}} \leq \left|p_L(\frac{2\pi}{3})\right|^2 \leq 3^{L-1}.$$

(Note that Proposition 3.5 was proved with the assumption that $|p(\omega)|^2 = P_L\left(\sin^2\left(\frac{\omega}{2}\right)\right)$. A review of the proof shows that the result holds when $p(\omega) = P_L\left(\sin^2\left(\frac{\omega}{2}\right)\right)$. In this case a power of 2 appears and the product grows twice as fast.)

This allows us to make an estimate about the decay of $\hat{\tilde{\varphi}}_N(\omega)$ when $|\omega|$ tends to $+\infty$. For example, if we let $\frac{\log 3}{\log 2} \approx 1.6$, then the decay is estimated by $|\omega|^{N+1-0.4L}$. To be sure, this is an asymptotic result for large values of L; however, it does demonstrate the possibility to have the desired decay for $\hat{\tilde{\varphi}}_N(\omega)$. This also implies that to meet the minimum requirements, it will be necessary to take $L \approx 2.5N$ for large N.

This asymptotic estimate thus gives information about the size of the filters. Indeed, it implies that if the filter $m_{0,N}$ has $N+1$ coefficients, then the size of $\tilde{m}_{0,N}$ needs to be about $9N$. We note, however, that examples calculated for small values of N require fewer coefficients than are indicated by this asymptotic estimate.

On the other hand, we note that the functions

$$\hat{\varphi}_N(\omega) = e^{-i\varepsilon_N \frac{\omega}{2}} \left(\frac{\sin\frac{\omega}{2}}{\frac{\omega}{2}}\right)^{N+1}$$

($\varepsilon_N = 0$ or 1) and $\hat{\psi}_N(\omega)$ become very localized about the origin when N becomes large. When this happens, the Riesz constants A

and $\tilde{A}$, as computed in Section 4.3, become very poor, and this in turn threatens the stability of the subdivision algorithms.

We summarize below the three cases $N = 1, 2$, and 3 and present graphs of the functions φ_N, $\tilde{\varphi}_N$, ψ_N and $\tilde{\psi}_N$. The filters $\tilde{m}_{0,N}$ have been chosen to have the minimal length consistent with the required decay for $\widehat{\tilde{\phi}}_N$. For this reason the functions $\tilde{\varphi}_N$ and $\tilde{\psi}_N$ are not very smooth.

This last point raises a comment. If one wishes to do approximations in a space of spline functions, then the family $\{\tilde{\psi}_{j,k}\}$ will be used for the analysis while the family $\{\psi_{j,k}\}$ will be used for the synthesis. Here, the lack of regularity of the $\tilde{\psi}_{j,k}$ is not a problem. On the other hand, where we need smoothness for synthesis, we know that ψ_N is by construction in $\mathcal{C}^{N-1}$.

The property that is desired for the analyzing wavelet $\tilde{\psi}_N$ is oscillation, which is to say that a number of moments vanish. But we know from the definition of $\tilde{\psi}_N$ (4.18) and the properties of the filters that this number is $N + 1$. Thus the biorthogonal systems provide a good balance between the properties that are desired for analysis and synthesis. Furthermore, the functions φ_N and $\tilde{\varphi}_N$ are symmetric (about 0 when N is odd and about $1/2$ when N is even), which corresponds to the filters having linear phase. The wavelets ψ_N and $\tilde{\psi}_N$ are symmetric when N is odd and anti-symmetric when N is even. An added attraction is that the filter coefficients have finite binary expansions. All of this justifies envisioning a broad range of applications to signal and image processing for these constructions.

Other spline wavelets have been constructed by Chui and Wang (1991), who followed a different point of view. They constructed a Riesz basis for W_0 while maintaining orthogonality between the scales. As a consequence, the dual wavelets are also spline functions, but they cannot have compact support.

As a small appendix to this chapter, we return to the issue of interpolation in V_0^N. Briefly, we wish to show that there is a unique function I in V_0^N such that $I(n) = \delta_{0,n}$ for $n \in \mathbb{N}$ when N is odd and that this cannot happen when N is even. We proceed formally by assuming that such an I exists. Then $I(x) = \sum \alpha_k \varphi(x-k)$ and $\hat{I}(\omega) = \alpha(\omega)\hat{\varphi}(\omega)$ where $\alpha(\omega) = \sum \alpha_k e^{-ik\omega}$ and $\alpha \in L^2[0, 2\pi]$. The condition $I(n) = \delta_{0,n}$ is equivalent to saying that

$$\sum_{l \in \mathbb{Z}} \hat{I}(\omega + 2l\pi) = 1$$

EXAMPLES AND APPLICATIONS   159

for all ω. Thus, if all of this is going to happen, we must have

$$\alpha(\omega) = \left[\sum_{l\in\mathbb{Z}} \hat\varphi(\omega+2l\pi)\right]^{-1}$$

and $\hat{I}(\omega) = [\sum \hat\varphi(\omega+2l\pi)]^{-1} \hat\varphi(\omega)$.

For the case at hand, $\hat\varphi(\omega) = e^{-i\varepsilon_N \frac{\omega}{2}} \left(\frac{\sin\frac{\omega}{2}}{\frac{\omega}{2}}\right)^{N+1}$ ($\varepsilon_N = 0$ or 1) and

$$\hat\varphi(\omega+2l\pi) = \hat\varphi(\omega)\left(\frac{\omega}{\omega+2l\pi}\right)^{N+1}.$$

Thus for $\omega \in [0, 2\pi[$,

$$\sum_{l\in\mathbb{Z}} \hat\varphi(\omega+2k\pi) = \hat\varphi(\omega) \sum_{l\in\mathbb{Z}} \left(\frac{\omega}{\omega+2k\pi}\right)^{N+1}. \qquad (4.157)$$

If N is odd, the terms in the sum on the right-hand side are all positive, and the function $\beta(\omega) = \sum \hat\varphi(\omega+2l\pi)$ does not vanish on $[0, 2\pi[$. If N is even, the sum on the right has a zero at $\omega = \pi$ because, in this case, the terms where $l = n$ and $l = -n-1$ cancel each other for all $n \in \mathbb{N}$.

In a neighborhood of $\omega = \pi$, we can express β as

$$\beta(\omega) = \hat\varphi(\omega)(\omega)^{N+1}\gamma(\omega)$$

where

$$\gamma(\omega) = \sum_{l\in\mathbb{Z}} \left(\frac{1}{\omega+2l\pi}\right)^{N+1}$$

is analytic. Since $\gamma'(\pi) = -(N+1)\sum\left(\frac{1}{\pi+2l\pi}\right)^{N+2}$ does not vanish (N is even), we conclude that β has a simple zero at $\omega = \pi$. Thus we can express β in a neighborhood of $\omega = \pi$ as $\beta(\omega) = (\omega-\pi)\eta(\omega)$ where $|\eta(\omega)|$ is bounded away from zero. But this implies that β^{-1} cannot be in $L^2[0, 2\pi]$ when N is even. Thus the assumption that I exists, which means that α must be in $L^2[0, 2\pi]$, leads to a contradiction in case N is even.

This discussion of the zero of β should not obscure the fact that β is a trigonometric polynomial. Indeed, $\beta(\omega) = \sum \varphi(n) e^{-in\omega}$, once again by the Poisson summation theorem. The sum on the right is finite since $\varphi(n) = 0$ for all but a finite number of n.

For the case N is odd, it is clear from the definition of $\hat\varphi$ and (4.157) that β is a continuous, 2π-periodic function that is bounded away from zero. This means that the obvious definition for I, which is $\hat{I}(\omega) = \beta(\omega)^{-1}\,\hat{\varphi}(\omega)$, makes sense, and this proves the existence of a function having the required properties. The uniqueness of I comes directly from the construction.

We end with a few comments about the mapping $TV_0^N \mapsto l^2$ for N odd defined by $Tf = \{f(k)\}$. The first thing to observe is that I and its integer translates form a Riesz basis. This follows directly from the fact that the family $\{\varphi(x-k)\}_{k\in\mathbb{Z}}$ is a Riesz basis and the properties of β. In particular, we know that there are two strictly positive constants C_1 and C_2 such that

$$C_1 \leq \sum_{l\in\mathbb{Z}} |\hat{\varphi}(\omega + 2l\pi)|^2 \leq C_2\,.$$

If we write $M = \max|\beta(\omega)|^2$ and $m = \min|\beta(\omega)|^2$ then this last inequality and the relation $\hat{\varphi}(\omega) = \beta(\omega)\hat{I}(\omega)$ shows that

$$\frac{C_1}{M} \leq \sum_{l\in\mathbb{Z}} |\hat{I}(\omega + 2l\pi)|^2 \leq \frac{C_2}{m}\,, \qquad (4.158)$$

which is equivalent to saying that the family $\{I(x-k)\}_{k\in\mathbb{Z}}$ is a Riesz basis for the space it spans. But this space is exactly V_0^N. To see this, assume that $f(x) = \sum f_k\,\varphi(x-k)$. Then

$$\hat{f}(\omega) = F(\omega)\hat{\varphi}(\omega) = \beta(\omega)F(\omega)\hat{I}(\omega)\,,$$

and $\beta F \in L^2[0, 2\pi]$. This proves that the function f is in the span of $\{I(x-k)\}_{k\in\mathbb{Z}}$.

In the space domain, (4.158) becomes

$$\frac{C_1}{M} \sum_{k\in\mathbb{Z}} |f(k)|^2 \leq \|f\|^2 \leq \frac{C_2}{m} \sum_{k\in\mathbb{Z}} |f(k)|^2\,, \qquad (4.159)$$

for all $f(x) = \sum f(k)I(x-k)$. Thus we have shown that the mapping $T^{-1}l^2 \mapsto V_0^N$ is 1-to-1, continuous and onto. It follows that T is also bijective and continuous.

We have proceeded formally without comment about the functions' membership in the various functional spaces, but it is indeed easy to see from the properties of φ and $\hat{\varphi}$ for $N \geq 1$ that all of the functions are well behaved and in the proper spaces so that all the formulas make sense.

As a final note to the chapter, we mention again that our main point of view has been 'projection,' and for this the natural basis is $\{\varphi(x-k)\}_{k\in\mathbb{Z}}$. On the other hand, if one takes the 'interpolation'

EXAMPLES AND APPLICATIONS

Table 4.2 *Summary of spline filters*

$N = 1$			
$m_{0,N}(\omega) = \cos^2(\frac{\omega}{2})$			
$h_0 = \frac{1}{2}$	$h_{\pm 1} = \frac{1}{4}$		
$\tilde{m}_{0,N}(\omega) = \cos^2(\frac{\omega}{2})P_2(\sin^2(\frac{\omega}{2}))$			
$\tilde{h}_0 = \frac{3}{4}$	$\tilde{h}_{\pm 1} = \frac{1}{4}$	$\tilde{h}_{\pm 2} = -\frac{1}{8}$	
$N = 2$	$A = 2^9$		
$m_{0,N}(\omega) = e^{-i\frac{\omega}{2}}\cos^3(\frac{\omega}{2})$			
$h_{-1} = \frac{1}{8}$	$h_0 = \frac{3}{8}$	$h_1 = \frac{3}{8}$	$h_2 = \frac{1}{8}$
$\tilde{m}_{0,N}(\omega) = e^{-i\frac{\omega}{2}}\cos^5(\frac{\omega}{2})P_4(\sin^2(\frac{\omega}{2}))$			
$\tilde{h}_0, \tilde{h}_1 = \frac{350}{A}$	$\tilde{h}_{-1}, \tilde{h}_2 = -\frac{26}{A}$	$\tilde{h}_{-2}, \tilde{h}_3 = -\frac{97}{A}$	
$\tilde{h}_{-3}, \tilde{h}_4 = \frac{19}{A}$	$\tilde{h}_{-4}, \tilde{h}_5 = \frac{15}{A}$	$\tilde{h}_{-5}, \tilde{h}_6 = -\frac{5}{A}$	
$N = 3$	$B = 2^{16}$		
$m_{0,N}(\omega) = \cos^4(\frac{\omega}{2})$			
$h_0 = \frac{3}{8}$	$h_{\pm 1} = \frac{1}{4}$	$h_{\pm 2} = \frac{1}{16}$	
$\tilde{m}_{0,N}(\omega) = \cos^8(\frac{\omega}{2})P_6(\sin^2(\frac{\omega}{2}))$			
$\tilde{h}_0 = \frac{64680}{B}$	$\tilde{h}_{\pm 1} = \frac{22638}{B}$	$\tilde{h}_{\pm 2} = -\frac{22438}{B}$	$\tilde{h}_{\pm 3} = -\frac{7228}{B}$
$\tilde{h}_{\pm 4} = \frac{83688}{B}$	$\tilde{h}_{\pm 5} = \frac{820}{B}$	$\tilde{h}_{\pm 6} = -\frac{2128}{B}$	$\tilde{h}_{\pm 7} = \frac{217}{B}$
$\tilde{h}_{\pm 8} = \frac{252}{B}$	$\tilde{h}_{\pm 9} = \frac{53}{B}$		

point of view, the natural basis is $\{I(x-k)\}_{k\in\mathbb{Z}}$. What we have done is to show the relation between these two bases.

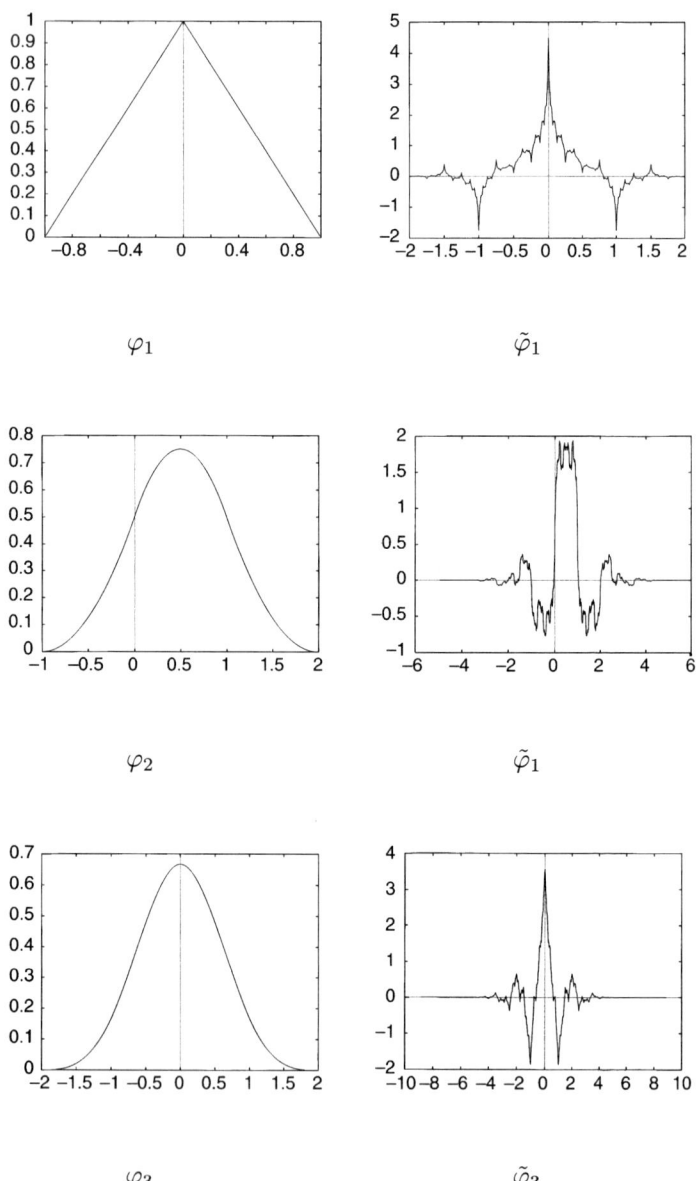

Figure 4.3

EXAMPLES AND APPLICATIONS

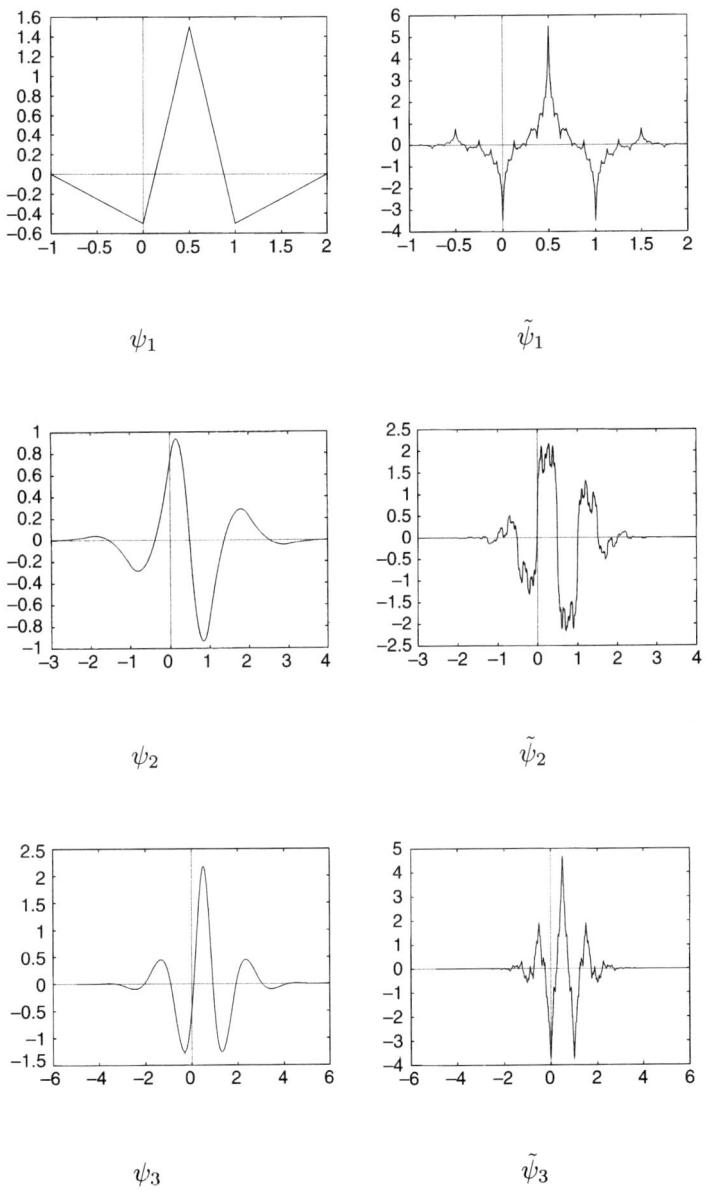

Figure 4.4

CHAPTER 5

Stochastic processes: multiscale processing and applications to signal and image compression

5.1 Introduction

Multiscale decompositions are *non-parametric* in the sense that they do not require any specific assumptions about the data to be decomposed. We require only that these data lie in a reasonable function or sequence space. So far, we have studied the different aspects of multiresolution approximation and wavelets without considering the properties of the functions or signals that one wishes to process in practical applications. We now want to check the effect of these properties on the resulting multiscale decompositions. Specifically, we wish to address the following kinds of questions:

- How well can a signal of a certain type be approximated in the spaces V_j? This is typically a linear approximation process, since this approximation can be performed by the projection operator.

- How well can a signal of a certain type be approximated by a combination of N wavelets? This, in contrast, is a non-linear approximation process, because we allow the choice of these wavelets to depend on the signal.

- Are these approximations better than those that are obtained if we replace the wavelet basis by the trigonometric system?

We formalize these questions as follows: If H is a Hilbert space and $\{e_k\}_{k>0}$ is an orthonormal basis for H, we define, for all $x \in H$ and $N > 0$, the linear approximation of x by

$$\mathcal{L}_N x = \sum_{k \leq N} \langle x, e_k \rangle e_k$$

and the non-linear approximation of x by

$$\mathcal{A}_N x = \sum_{k \in E_N} \langle x, e_k \rangle e_k,$$

where $E_N = E_N(x)$ represents the set of indices corresponding to the N largest coordinates of x, i.e., $\mathrm{Card}(E_N) = N$ and

$$k \in E_N, l \notin E_N \Rightarrow |\langle x, e_l \rangle| \leq |\langle x, e_k \rangle|.$$

We wish to evaluate the errors $\|x - \mathcal{L}_N x\|_H$ and $\|x - \mathcal{A}_N x\|_H$ when H is the space $L^2(I)$, $\{e_n\}_{n \in \mathbb{Z}}$ is either a wavelet or trigonometric basis, and x belongs to a certain class of signals. Here I can be an interval $[a, b]$ in the case of univariate signals or a rectangle $[a, b] \times [c, d]$ in the case of images.

From a practical point of view, these questions are related to the problem of data compression and, in particular, to image compression.

Since the 1970s and 1980s, the problem of data compression has become a central issue both in information theory and in signal processing. Signals such as still or animated images, speech, or music, in their digital form, are represented by a considerable number of bits, which must, typically, be stored, transmitted, or processed. The goal of data compression is to reduce significantly the number of bits that must be handled, while keeping the essential information about the signal that will be necessary for a given application. Note that for a continuous (or analog) signal, this original amount of data is infinite. In the case of digitized signals, one can define the 'compression ratio' as N_0/N_1 where N_0 and N_1 represent respectively the number of bits in the signal before and after compression. In the case of digitized signals, one classically distinguishes two types of compression:

• Lossless compression, which results from using algorithms such as Huffman or Lempel–Ziv coding. In these cases, a sequence of bits is transformed into a smaller one by an invertible operation. The signal can thus be recovered without error.

• Lossy compression, which results from using an algorithm that reduces the number of bits and recovers an approximation of the original signal. In this case, there is no theoretical limit to the compression ratio; however, this ratio must be chosen so that the approximation error is acceptable for the given application.

INTRODUCTION

One often uses a lossy algorithm to compress the signal; this is then followed by a lossless coding algorithm for transmission. In this chapter, we will be concerned only with lossy compression techniques.

Over the last decade, multiscale methods — specifically, wavelet bases — have become popular in image processing, particularly for compression. This popularity is because they yield sparse decompositions of the function $I(x,y)$ that represents the light intensity of the image. Here 'sparse' means that a small number of coefficients carry most of the significant information. This property can be described more precisely in terms of non-linear wavelet approximation. It is important to note, however, that compressing an image is a more complex process than simply approximating it with a small number of coefficients. Compression also involves the quantization of these coefficients, which means encoding them with a finite number of bits. The problem of choosing an optimal quantization strategy is beyond the scope of this book, and we refer to Antonini, Barlaud, Daubechies, and Mathieu (1992) for a complete account.

In the case of wavelet bases, an important class of results relates the approximation of a function f to the smoothness of f. For example, assuming that the wavelet ψ has M vanishing moments, then, for all $s \in \,]0, M]$,

$$f \in \Lambda^s(I) \Leftrightarrow \|f - \mathcal{L}_N f\|_{L^\infty} \leq CN^{-s}.$$

Here, $\Lambda^s(I)$ is the standard Hölder space $\mathcal{C}^s(I)$ if s is not an integer. $\Lambda^n(I)$ is defined by

$$\left\{ f \in \mathcal{C}^{n-1} \mid |f^{(n-1)}(t-h) + f^{(n-1)}(t+h) - 2f^{(n-1)}(t)| \leq C|h|^2 \right\}$$

if n is an integer. For Sobolev regularity, we have $f \in H^s(I)$ if and only if

$$\sum_{N \geq 0} N^{2s-1} \|f - \mathcal{L}_N f\|_{L^2}^2 < +\infty \Rightarrow \|f - \mathcal{L}_N f\|_{L^2} \leq CN^{-s}.$$

These results about linear wavelet approximation are detailed in Daubechies (1992) and Meyer (1990).

Similar results have been obtained by Ronald DeVore, Bjorn Jawerth, and Vasil Popov (1992) for non-linear approximation in the general framework of Besov spaces. DeVore, Jawerth, and Lucier (1992) have used these results for wavelet-based image compression algorithms. They note that real images, although they present singularities related to edges and texture, have some regularity in the sense of certain Besov norms.

In this chapter, we take a different track. The signals we study will be described as random functions $f(t,\omega)$, and their properties will be expressed in probabilistic terms. The precision of the approximations will be measured by the mean square errors $E(\,\|f - \mathcal{L}_N f\|_2^2\,)$ and $E(\,\|f - \mathcal{A}_N f\|_2^2\,)$.

The introduction of randomness to describe images and the use the mean square error reflect the realities of image compression. In practice, one expects a compression algorithm to give good results on most of the images, with the realization that it may be less efficient on some pathological images that may appear. Because of its complexity, it is impractical to have a full description of the probability distribution of f. For example, in the case of a digital image of size 256×256, this is a distribution in $\mathbb{R}^{2^{16}}$. However, one can easily access partial information about this distribution, and this information is typically presented in terms of expectation functions. We will use this information to study the expected error for two different bases: the trigonometric system and wavelet bases.

Note that the choice of an L^2 error criterion is quite arbitrary. In the case of images, it appears that this L^2 error does not match the error measurement made by the human visual system, which seems to be more sensitive in many situations. Although other norms have been proposed, we will keep the L^2 norm. In doing so, we note that, in the case of wavelet decompositions, all the error computations that we present can be generalized to other norms, using the property that wavelet bases are unconditional bases for most function spaces (see Meyer (1990) and Daubechies (1992)). A remark, which is related to the choice of a norm, is that it is not clear that the eye uses a norm to measure the error at the low level stage of visual processing. A norm that fits ideally with the eye's measurement of error should take into account both the regularity of the image in the regions representing smooth objects and the existence of discontinuities representing the edges of these objects.

The rest of this chapter is organized as follows: In Section 5.2, we review linear approximation results that were obtained in the case where the only available information is given by the first and second order moments of the distribution. In this case, the performance of Fourier series and of wavelet bases on images are comparable. We apply these results to images in Section 5.3, and we present the results of actual computations in Section 5.4. A more sophis-

LINEAR APPROXIMATION

ticated model to describe images is proposed in Section 5.5. This model leads to non-linear approximation results that are detailed in Section 5.6. In that case, we show that a well-chosen wavelet basis outperforms Fourier series and any type of linear approximation. Examples of non-linear approximations of real images are also presented in Section 5.4.

For the sake of simplicity, we will establish our results by considering unidimensional signals that are described by a stochastic process $f(t, \omega)$. We then describe how these results can be generalized to multidimensional signals and applied to images.

5.2 Linear approximation

Let $s(t)$ be a stochastic process of the second order defined on $[0, 1]$, i.e., $E(|s(t)|)$ and $E(|s(t)|^2)$ are bounded functions. (We refer to Oppenheim and Schafer (1975) for basic material about stochastic processes.) In this section, we assume that the only information we have about the process $s(t)$ is given by its autocorrelation function

$$R(t_1, t_2) = E(s(t_1)\overline{s(t_2)}). \tag{5.1}$$

A natural question is how can we use this information to compare the effects of the decompositions of a signal in different bases? Given an orthonormal basis $\{e_n\}_{n \geq 0}$, it is possible to compute the mean square value of the coordinates of $s(t)$ since we have

$$\begin{aligned}
E(|\langle s, e_n \rangle|^2) &= E\Big(\Big|\int_0^1 s(t)\overline{e_n(t)}\,dt\Big|^2\Big) \\
&= E\Big(\iint_{[0,1]^2} s(t)\overline{s(u)}\,\overline{e_n(t)}e_n(u)\,dt\,du\Big) \\
&= \iint_{[0,1]^2} R(t,u)\overline{e_n(t)}e_n(u)\,dt\,du \\
&= \langle R, e_{n,n} \rangle,
\end{aligned}$$

with $e_{m,n}(x,y) = e_m(x)\overline{e_n(y)}$. Thus, the autocorrelation function allows us to estimate the mean square error between $s(t)$ and the linear approximation of $s(t)$ by its first N coordinates,

$$s_N(t) = \sum_{n=0}^{N-1} \langle s, e_n \rangle e_n(t), \tag{5.2}$$

since
$$\varepsilon(N) = E(\|s - s_N\|^2) = \sum_{n \geq N} E(|\langle s, e_n \rangle|^2). \tag{5.3}$$

The basis that minimizes this quantity for all N is the Karhunen–Loève basis. This is the basis consisting of the orthonormal eigenfunctions $\{e_n\}_{n \geq 0}$ of the integral operator

$$\mathcal{R}f(t) = \int_0^1 R(t, u) f(u) \, du, \tag{5.4}$$

rearranged so that the corresponding eigenvalues $\lambda_n \geq 0$ are decreasing with n.

In most cases, these eigenfunctions are not explicitly available, and consequently the computation of the coefficients of a discretized function f in the Karhunen–Loève basis requires a large number of operations. Fortunately, it is often possible to obtain a near optimal approximation with a simpler system that is better adapted for numerical computations.

We first consider the case of an orthonormal wavelet basis. For this, we must mention that the bases we have constructed for the real line can be adapted to the interval (see Cohen, Daubechies, and Vial (1993)). $L^2[0,1]$ is approximated by a multiresolution analysis, which, in this case, is a ladder of closed subspaces

$$V_{j_0} \subset V_{j_0+1} \subset V_{j_0+2} \cdots \to L^2[0,1], \tag{5.5}$$

with $j_0 \geq 0$. V_j is generated by 2^j orthonormal scaling functions $\varphi_{j,k}$, $k = 0, \ldots, 2^j - 1$, with the property that the support of $\varphi_{j,k}$ is in the interval $[2^{-j}(k-c), 2^{-j}(k+c)]$, and c does not depend on j. At each level, the orthonormal complement of V_j in V_{j+1} is generated by 2^j orthonormal wavelets $\psi_{j,k}$, $k = 0, \ldots, 2^j - 1$, such that $\operatorname{supp}(\psi_{j,k}) \subset [2^{-j}(k-c), 2^{-j}(k+c)]$. As a consequence, the family

$$\bigcup_{j \geq j_0} \{\psi_{j,k}\}_{k=0,\ldots,2^j-1}, \tag{5.6}$$

completed by $\{\varphi_{j_0,k}\}_{k=0,\ldots,2^{j_0}-1}$, constitutes an orthonormal basis of $L^2[0,1]$. Although we do not make use of this property, we mention that when the support of the scaling functions or the wavelets do not contain 0 or 1, these functions are simply defined in the usual way from the standard scaling function and wavelet φ and ψ with the notation $f_{j,k} = 2^{j/2} f(2^j \cdot - k)$.

Results about linear approximation in a wavelet basis relate the regularity of $R(t,u)$ on the diagonal line $\{(t,u)\,|\,t=u\}$ and the cancellation properties of the wavelets, that is, their number of vanishing moments. However, equivalent results can be stated in terms of linear approximation in the spaces V_j. In this case, the functions $\psi_{j,k}$ do not appear, and one considers the degree to which the scaling functions $\varphi_{j,k}$ reproduce polynomials.

Here we give the formulation using the cancellation properties of the wavelets. For this, we reorder the wavelet basis (5.6) by defining $e_n = \varphi_{j_0,n}$ when $n < 2^{j_0}$ and $e_n = \psi_{j,k}$, when $n = 2^j + k$, $0 \le k < 2^j$, $j \ge j_0$.

For $\alpha > 0$, a real function $F(x)$, $x = (x_1, \ldots, x_d) \in \mathbb{R}^d$, is said to be $\mathcal{C}^\alpha$ at $z \in \mathbb{R}^d$ if and only if there exists a polynomial

$$P(x) = \sum_{|m|<\alpha} a_m\, x^m$$

(with $|m| = m_1 + \cdots + m_d$ and $x^m = x_1^{m_1} \cdots x_d^{m_d}$), such that for all x in a neighborhood V_z of z,

$$|P(x) - F(x)| \le C\|x - z\|^\alpha. \tag{5.7}$$

Note that for $\alpha \in \mathbb{N}$ this is weaker than α differentiability.

Theorem 5.1 *Suppose that $R(t,u)$ is $\mathcal{C}^\alpha$ at all points of the diagonal line $\{(t,t)|t \in [0,1]\}$ for a fixed $\alpha > 0$ and that for all $m \in \mathbb{N} \cap [0,\alpha[$, $j \ge j_0$ and $k = 0, \ldots, 2^j - 1$, $\int_0^1 x^m \psi_{j,k}(x)dx = 0$. Then*

$$\varepsilon(N) \le CN^{-\alpha}. \tag{5.8}$$

Proof. It is clearly sufficient to prove (5.8) for $N = 2^p$, $p \ge j_0$. By a compactness argument, the property that $R(t,u)$ is $\mathcal{C}^\alpha$ along the diagonal means that there exists $C > 0$ such that for each $v \in [0,1]$, there is a polynomial $P_v(t,u)$ of global degree strictly less than α with

$$|R(t,u) - P_v(t,u)| \le C(|t-v| + |u-v|)^\alpha \tag{5.9}$$

for all $(t,u) \in [0,1]^2$. Using this estimate at $v = 2^{-j}k$, we get

$$E(|\langle s, \psi_{j,k}\rangle|^2) = \iint_{[0,1]^2} R(t,u)\psi_{j,k}(t)\psi_{j,k}(u)\,dt\,du$$

$$= \iint_{[0,1]^2} (R(t,u) - P_{2^{-j}k}(t,u))\overline{\psi_{j,k}(t)}\psi_{j,k}(u)\,dt\,du$$

$$\le C \iint_{[0,1]^2} (|t-v| + |u-v|)^\alpha |\psi_{j,k}(t)\psi_{j,k}(u)|\,dt\,du.$$

From the support properties of the functions $\psi_{j,k}$ and the Schwarz inequality, we derive

$$\begin{aligned} E(|\langle s, \psi_{j,k}\rangle|^2) &\leq C(2c)^\alpha 2^{-\alpha j} \iint_{[0,1]^2} |\psi_{j,k}(t)\psi_{j,k}(u)|\,dt\,du \\ &\leq C(2c)^{\alpha+1} 2^{-(\alpha+1)j}. \end{aligned}$$

Summing on all $k = 0, \ldots, 2^j - 1$, then on all $j \geq p$, we finally obtain the desired estimate, which proves the theorem. □

REMARK

This result is in fact a simple rephrasing, in the stochastic framework, of the deterministic results on the multiresolution approximation of functions in Sobolev spaces that we stated in the introduction of this chapter: $\alpha/2$ represents the degree of differentiability in the mean square sense of the process $s(t)$.

We now consider the trigonometric system, $e_n(t) = e^{i2\pi nt}$, where $n \in \mathbb{Z}$. In this case, an estimate on the quantity

$$E(|\langle s, e_n\rangle|^2) = \iint_{[0,1]^2} R(t,u) e^{i2\pi n(u-t)}\,dt\,du \qquad (5.10)$$

cannot be easily derived from the regularity of $R(t, u)$ unless R is the restriction of a regular, $\mathbb{Z}^2$-periodic function. Most signals, and in particular images, do not satisfy this property. Thus, to estimate $\varepsilon(N)$, one needs more information about $R(t, u)$.

We note that an important class of signals in one dimension satisfies the stationarity property, which means that $R(t, u) = r(|t-u|)$. In case $r(t)$ is a 1-periodic function, the Karhunen–Loève basis is given by the trigonometric system. As mentioned, the signals that we have in mind do not satisfy this property. Specifically, an image, a video sequence, or a piece of speech is the restriction to a finite domain of a non-periodic function. For these signals, $r(t)$ is typically an even function that decreases on $[0, +\infty[$.

5.3 Linear approximation of images

For TV images, a commonly used model for the autocorrelation function $E(I(x,y)I(x',y'))$ is

$$R(x, x', y, y') = R_0 e^{-\mu(|x-x'|+|y-y'|)}, \qquad (5.11)$$

where R_0 and μ are constants that depend on the normalization of the light intensity function and on the size of the image (see for example Kak and Rosenfeld (1982)). In particular, $I(x,y)$ is assumed to be centered around zero. It takes the value $I_{\max}$ in the white region and $-I_{\max}$ in the black region. With this in mind, we will focus on the one-dimensional stationary processes whose autocorrelation function is

$$R(t,u) = e^{-|t-u|}. \qquad (5.12)$$

In this case, Theorem 5.1 indicates that the linear approximation error $\varepsilon(N)$ with a wavelet basis is dominated by CN^{-1}.

For the trigonometric system, we see that

$$\begin{aligned}
E(|\langle s, e_n \rangle|^2) &= \iint_{[0,1]^2} e^{-|t-u|} e^{i2\pi n(u-t)} \, dt \, du \\
&= \int_{-1}^{1} e^{-|t|} (1-|t|) e^{i2\pi n t} \, dt \\
&= \int_{0}^{1} (1-t) e^{-t} (e^{i2\pi n t} + e^{-i2\pi n t}) \, dt \\
&= \frac{2}{1+4\pi^2 n^2} + \frac{(8\pi^2 n^2 - 2)(1 - 1/e)}{(1+4\pi^2 n^2)^2}.
\end{aligned}$$

As N goes to infinity, we obtain the estimates

$$\varepsilon(N) = \sum_{|n| \geq N/2} E(|\langle s, e_n \rangle|^2) + O(N^{-2}),$$

and

$$\sum_{|n| \geq N/2} E(|\langle s, e_n \rangle|^2) = \frac{2(2 - 1/e)}{\pi^2 N} + O(N^{-2}), \qquad (5.13)$$

which shows that the trigonometric system performs as well as a wavelet basis (for these particular processes). In fact, all these systems are near optimal for linear approximation, in the sense that they perform as well as the Karhunen–Loève basis. This can be checked by determining explicitly the Karhunen–Loève functions for this case. To see this, assume that $\lambda > 0$ is a given eigenvalue of $\mathcal{R}$ (defined by (5.4)). Then the associated eigenfunction is C^{∞} and satisfies the equation

$$\begin{aligned}
\lambda f(t) &= \int_{0}^{1} e^{-|t-u|} f(u) \, du \\
&= e^{-t} \int_{0}^{t} e^{u} f(u) \, du + e^{t} \int_{t}^{1} e^{-u} f(u) \, du.
\end{aligned}$$

Differentiating once, we obtain

$$\lambda f'(t) = -e^{-t} \int_0^t e^u f(u)\, du + e^t \int_t^1 e^{-u} f(u)\, du, \qquad (5.14)$$

which shows that necessarily

$$f'(0) = f(0) \quad \text{and} \quad f'(1) = -f(1). \qquad (5.15)$$

After a second differentiation, we obtain the equation

$$\lambda f''(t) = (\lambda - 2) f(t). \qquad (5.16)$$

It is clear that $\|\mathcal{R}\| \leq 1$ so $(\lambda - 2)/\lambda \leq 0$. The solutions of (5.16) are thus of the form $a\cos(\omega t) + b\sin(\omega t)$ with $\omega = \sqrt{2/\lambda - 1}$. From the boundary conditions (5.15), we finally obtain the family of orthogonal eigenfunctions

$$e_n(t) = \pi n \cos(\pi n t) + \sin(\pi n t), \quad n \in \mathbb{Z}/\{0\}, \qquad (5.17)$$

with the associated eigenvalues

$$\lambda_n = (1 + \pi^2 n^2)^{-1}. \qquad (5.18)$$

It follows that the approximation error in the Karhunen–Loève basis satisfies

$$\varepsilon(N) = 2(\pi^2 N)^{-1} + O(N^{-2}). \qquad (5.19)$$

The conclusion is that, when the autocorrelation is given by (5.12), the asymptotic performances of the Karhunen–Loève basis, the trigonometric system, and wavelet bases are equivalent for linear approximation.

The analogous estimates for the bidimensional case (5.11) follow immediately because the function $R(x, x', y, y')$ is separable. Consequently, the linear approximation error in V_j is of order 2^{-j}, or equivalently $\varepsilon(N)$ is of order $N^{-1/2}$ for the tensor product wavelets. The optimal linear approximation error (for the Karhunen–Loève basis) is of order $(\log N)^2 N^{-1}$. This can be achieved by using a fully separable wavelet basis $\Psi_{j,j',k,k'}(x,y) = \psi_{j,k}(x)\psi_{j',k'}(y)$, $j, j' \geq 0$, $0 \leq k \leq 2^j - 1$, $0 \leq k' \leq 2^{j'} - 1$, rearranged in increasing order of $j + j'$.

5.4 Approximation and compression of real images

These results allow one to predict the performance of a simplistic compression strategy that consists in encoding an approximation of the image while discarding the details at finer scales. This strategy is clearly not visually optimal: For a given compression ratio, it is preferable to use some of the available bits to encode certain fine-scale details that are visually important, particularly edges.

From a theoretical point of view, the previous results indicate that the approximation error should decay like 2^{-j}; this is confirmed by experiment. Figure 5.1 represents the original 256×256 pixels image that is being processed. (All figures are at the end of the chapter.) Figures 5.3, 5.4, 5.5, and 5.6 show, respectively, the performance of linear approximation using the Haar system, a more regular orthonormal wavelet associated with a filter having four coefficients ($h_0 = -0.1$, $h_1 = 0.2$, $h_2 = 0.6$, $h_3 = 0.3$, and $\varphi, \psi \in C^{0.5}$), a biorthogonal wavelet basis associated with the linear spline approximation (see Section 4.5.2), and the trigonometric system. The subfigures a, b, c, and d show, respectively, the approximations obtained by retaining the first 2^{2j} coefficients with $j = 4, 5, 6$, and 7.

In practice, we measure the signal-to-noise ratio defined by

$$\mathrm{SNR}(j) = \frac{10}{\log 10} \log\left(\frac{E}{\varepsilon(j)}\right),$$

where E is the L^2 norm of the original image and $\varepsilon(j)$ the actual measured error at scale j. The signal-to-noise ratio is given in *decibels*, which is abbreviated db.

According to our theoretical results, this quantity must decrease linearly with j, and, indeed, this is what we observe:

- For the Haar system,
  SNR(7)=29db, SNR(6)=25db,
  SNR(5)=21db, SNR(4)=17db.
- For the orthonormal wavelet basis,
  SNR(7)=30db, SNR(6)=26db,
  SNR(5)=22db, SNR(4)=18db.
- For the biorthogonal wavelet basis,
  SNR(7)=30db, SNR(6)=26db,
  SNR(5)=22db, SNR(4)=18db.
- For the trigonometric system,
  SNR(7)=29db, SNR(6)=25db,
  SNR(5)=22db, SNR(4)=18db.

One should note that the average increment of 4db does not correspond exactly to a factor 2^{-j} but rather to 2.5^{-j}.

Note that the poor visual quality of the approximations with the Haar system is not reflected in the measured signal-to-noise ratio. This is because the criterion we use is based on the L^2 norm, and it does not take into account the regularity of the image.

Also note that blocking effects appear when we use the more regular wavelets at low scale approximation. These effects disappear when we use the trigonometric system, whose main defect is to introduce Gibbs phenomena. These appear here as oscillations near the edges. In contrast, as we mentioned in Chapter 4, these oscillations are limited in space when using approximation with compactly supported scaling functions.

Figure 5.2 represents the absolute value of the wavelet coefficients of our original image (in the Haar basis), the value zero being indicated by a black pixel. This image reveals the sparsity of the decomposition and the fact that numerically significant values of the wavelet coefficients persist at the finest scales, particularly in the vicinity of edges. This suggest that non-linear approximation, which means keeping the largest wavelet coefficients, should give better results than those obtained with linear approximation, for a given compression ratio.

In Figures 5.7, 5.8, 5.9, and 5.10 we have reconstructed the image using the 2^{12} largest coefficients of the decompositions in the four bases that were used above for linear approximations. Thus the number of retained coefficients is the same as for linear approximations with $j = 6$, and these figures should be compared with Figures 5.3.c, 5.4.c, 5.5.c, and 5.6.c

In the case of the Haar system, although the discontinuities of the generating functions create visual artifacts, we measure SNR=29db. This is 4db better than the linear approximation with the same number of coefficients. This image also reveals the consequence of the thresholding procedure that we have used: The resolution is better in areas where the light intensity varies strongly (particularly near edges) since we have kept the fine scale coefficients at these locations.

For the regular orthonormal wavelet basis SNR=30db, and for the biorthonormal wavelet basis we measure SNR=31db, which is even better. Clearly the visual results are more satisfactory.

Finally, we obtain SNR=26db for the trigonometric system. This seems to indicate that there is no significant improvement over the linear approximation where SNR=25db. Notice here that oscillations are created because high frequencies are added to the whole image.

These last experimental results suggest that we analyze the performance of non-linear approximations for the three different bases for a given stochastic model of the signals. So far, our description of signals and images has been limited to second-order statistical information. This model, although exact, is not rich enough to describe the fine structure of images or to be useful for non-linear approximation and compression. The problem, in the case of images, is that the autocorrelation function averages the smooth regions (which correspond to homogeneous objects) and the isolated discontinuities (which correspond to sharp edges).

To investigate non-linear approximation, we must introduce more information about the process $s(t)$. The model that we present in the next section is an attempt to describe the notion of a piecewise smooth stochastic processes.

5.5 Piecewise stationary processes

A class of processes designed to reflect the 'piecewise smooth' property in a stochastic framework was introduced by A. Cohen and J.P. d'Ales as follows.

A finite set of discontinuities $D = \{d_1, d_2, \ldots, d_L\} \subset [0, 1]$ with $d_i \leq d_{i+1}$ is obtained as the realization on $[0, 1]$ of a Poisson process with parameter $\mu > 0$. This means that the number of discontinuities is a random number with probability law

$$P(|D| = L) = e^{-\mu}\frac{\mu^L}{L!}, \tag{5.20}$$

and that, given the event $|D| = L$, the distribution of $(d_1, \ldots, d_L)$ is uniform over the simplex $\{0 \leq x_1 \leq \cdots \leq x_L \leq 1\}$.

Given the event $\{d_1, \ldots, d_L\}$, we set $d_0 = 0$ and $d_{L+1} = 1$, and we define $s(t)$ on $[d_i, d_{i+1}[$, $i = 0, \ldots, L$, by

$$s(t) = s_i(t), \tag{5.21}$$

where the functions s_i are independent realizations of a stationary process with autocorrelation function $R(t, u) = r(|t-u|)$ and mean M. We assume that $r(t)$ is twice differentiable, that is, the process that describes $s(t)$ between the discontinuities is differentiable in the mean square sense.

It is important to note that the global process $s(t)$ is also stationary. Its autocorrelation function is given by

$$R_s(t,u) = P(t,u)R(t,u) + (1 - P(t,u))M^2, \quad (5.22)$$

where $P(t,u) = e^{-\mu|t-u|}$ is the probability that no discontinuity d_i lies between t and u. We thus have

$$R_s(t,u) = M^2 + e^{-\mu|t-u|}(r(|t-u|) - M^2) = r_s(|t-u|). \quad (5.23)$$

Note that in the simplest case where $s_i(t) = s_i$ are independent realizations of a constant, centered process, the autocorrelation function is given by

$$R_s(t,u) = r_s(0)e^{-\mu|t-u|}. \quad (5.24)$$

This shows that the parameter μ in the model (5.11) can be interpreted as the Poisson density of discontinuities on a line in a real image.

REMARKS

• These processes are also good models for the evolution in time of the intensity of a fixed pixel in a video sequence. The discontinuities correspond to an edge in motion that crosses the pixel at a given time or to an abrupt change of the image.

• The generalization of these processes to model bidimensional signals is not straightforward. The discontinuities are curves rather than isolated points. So far, we have only investigated the unidimensional situation. Nevertheless, this case reveals the kinds of results that we can expect in the multidimensional setting.

We now present a result that shows that these piecewise smooth processes cannot be well approximated if one proceeds linearly.

Theorem 5.2 *Let λ_n, $n \geq 0$, be the sequence of eigenvalues of the integral operator $\mathcal{R}_s$ associated with the kernel $R_s(t,u)$. Assume that $\lambda_{n+1} \leq \lambda_n$ and that the eigenvalues are repeated according to their multiplicity. Then there exists a constant $C > 0$ such that, for all $N \geq 0$,*

$$\varepsilon(N) = \sum_{n \geq N} \lambda_n \geq CN^{-1}. \quad (5.25)$$

Consequently, the mean square error for any linear approximation of $s(t)$ cannot be less than CN^{-1}.

Proof. From (5.23), we can write

$$R_s(t,u) = K_A(t,u) + K_B(t,u), \quad (5.26)$$

where
$$K_A(t, u) = e^{-\mu|t-u|}(r(0) - M^2),\qquad (5.27)$$
and
$$K_B(t, u) = M^2 + e^{-\mu|t-u|}(r(|t-u|) - r(0)).\qquad (5.28)$$
Note that $K_B(t,u)$ belongs to $\mathcal{C}^2$. Denote the associated decomposition of the autocorrelation operator by
$$\mathcal{R}_s = A + B.\qquad (5.29)$$

It is clear that A is a positive operator. From the discussion in Section 5.4, we know that its eigenvalues $\{a_n\}_{n\geq 0}$, ordered as a decreasing sequence, satisfy
$$\lim_{n\to+\infty} n^2 a_n = C,\qquad (5.30)$$
where C depends on the parameters M, $r(0)$, and μ. In contrast, B is not necessarily a positive operator. We denote the singular values of B by $\{b_n\}_{n\geq 0}$. These are the eigenvalues of $|B| = (B^*B)^{1/2}$ ordered as a decreasing sequence. To prove (5.25), we will use the following result due to Ky Fan (see Gohberg and Krein (1969)):

Let A and B be compact operators in a Hilbert space and let $C = A + B$. Let a_n, b_n, and c_n be the associated sequences of singular values and assume that, for some $r > 0$,
$$\lim_{n\to+\infty} n^r a_n = C,\qquad (5.31)$$
and
$$\lim_{n\to+\infty} n^r b_n = 0.\qquad (5.32)$$
Then
$$\lim_{n\to+\infty} n^r c_n = C.\qquad (5.33)$$

We thus concentrate on proving that
$$\lim_{n\to+\infty} n^2 b_n = 0.\qquad (5.34)$$
By the Karhunen–Loève theorem, we know that
$$\sum_{n\geq N} b_n = \min_{\{e_n\}\text{ o.b.}} \sum_{n\geq N} \langle e_n, |B| e_n\rangle,\qquad (5.35)$$
where the minimum is taken over all orthonormal bases. Consider a

wavelet basis type (5.6), reordered as in Section 5.4, and such that the functions $\psi_{j,k}$ have two vanishing moments. Then for $p \geq j_0$,

$$\sum_{n \geq 2^p} b_n \leq \sum_{j \geq p, k} \langle \psi_{j,k}, |B|\psi_{j,k}\rangle. \tag{5.36}$$

Recall that there exists a unitary operator U such that $B = U|B|$, and define $\tilde{\psi}_{j,k} = U\psi_{j,k}$ where U is such an operator. Using the Schwarz inequality, we see that

$$\begin{aligned}
\langle \psi_{j,k}, |B|\psi_{j,k}\rangle &= \langle \tilde{\psi}_{j,k}, B\psi_{j,k}\rangle \\
&= \iint_{[0,1]^2} \overline{K_B(t,u)\psi_{j,k}(u)}\tilde{\psi}_{j,k}(t)\, dt\, du \\
&\leq \left(\int_0^1 \left|\int_0^1 K_B(t,u)\psi_{j,k}(u)du\right|^2 dt\right)^{1/2} \\
&\leq \sup_{t \in [0,1]} \left|\int_0^1 K_B(t,u)\psi_{j,k}(u)\, du\right|.
\end{aligned}$$

Since $K_B(t,u)$ is twice differentiable and the functions $\psi_{j,k}$ have two vanishing moments, we see that

$$\sup_{t \in [0,1]} \left|\int_0^1 K_B(t,u)\psi_{j,k}(u)\, du\right|$$
$$\leq \sup\left|\left(\frac{\partial}{\partial u}\right)^2 K_B\right| \int_0^1 \frac{|u - 2^{-j}k|^2}{2} |\psi_{j,k}(u)|\, du$$
$$\leq C\|\psi_{j,k}\|_{L_1} \int_{|2^j u - k| \leq c} \frac{|u - 2^{-j}k|^2}{2}\, du$$
$$\leq C 2^{-5j/2}.$$

Thus

$$\langle \psi_{j,k}, |B|\psi_{j,k}\rangle \leq C 2^{-5j/2}, \tag{5.37}$$

and

$$\sum_{n \geq N} b_n \leq C N^{-3/2}. \tag{5.38}$$

Finally, since the b_n are positive and decreasing, (5.38) implies that

$$b_n \leq C n^{-5/2}. \tag{5.39}$$

Thus Ky Fan's theorem applies, $\lim_{n \to +\infty} n^2 \lambda_n = C$, and (5.25) follows directly. □

NON-LINEAR APPROXIMATION

REMARK

A more natural idea for estimating the decay of b_n is to use the equivalent definition of the singular values of B that is given by

$$b_n = \inf_{T \in \mathcal{T}_n} \|B - T\|, \qquad (5.40)$$

where $\mathcal{T}_n$ is the space of operators with rank at most equal to n. It also seems natural to estimate this infimum by choosing a kernel K_T that is an approximation of K_B by a sum of n separable functions, typically its projection in a multiresolution approximation. However, one can check that this approach only leads to an estimate of order n^{-2}. Thus, in this case, an optimal approximation of the kernel does not lead to an optimal approximation of the operator.

5.6 Non-linear approximation

We now turn to the non-linear approximation

$$\mathcal{A}_N s = \sum_{k \in E_N} \langle s, e_k \rangle e_k, \qquad (5.41)$$

where $E_N = E_N(s)$ is the set of indices of the N largest coordinates of the process $s(t)$ that was described in Section 5.5. In this section, the mean square error will be defined by

$$\varepsilon(N) = E(\|\mathcal{A}_N s - s\|_2^2) = E\Big(\sum_{n \notin E_N(s)} |\langle s, e_n \rangle|^2 \Big). \qquad (5.42)$$

We first consider the case where $\{e_n\}_{n \geq 0}$ is a wavelet basis of the type (5.6), that is $\{\varphi_{j_0,k}\}_k \cup \{\psi_{j,k}\}_{j \geq j_0, k}$ reordered as in Section 5.4. The following theorem shows that non-linear wavelet approximation performs as well as if there were no discontinuities in the process $s(t)$.

Theorem 5.3 *Assume that $r(t)$ is in $\mathcal{C}^\alpha$ and $\int_0^1 x^m \psi_{j,k}(x) dx = 0$ for all $m \in \mathbb{N} \cap [0, \alpha[$, $j \geq j_0$, and $k = 0, \ldots, 2^j - 1$. Then*

$$\varepsilon(N) \leq C N^{-\alpha}. \qquad (5.43)$$

Proof. As in Theorem 5.1, it is clearly sufficient to prove (5.43) for $N = 2^p$ with $p \geq j_0$. Note that if we define

$$\mathcal{A}'_N s = \sum_{k \in E'_N(s)} \langle s, e_k \rangle e_k, \qquad (5.44)$$

where $E'_N(s) \subset \mathbb{N}$ has cardinal $|E'_N| \leq N$, we always have

$$\varepsilon(N) \leq \varepsilon'(N) = E(\|\mathcal{A}'_N s - s\|_2^2). \tag{5.45}$$

Based on this remark, it is sufficient to build a near-optimal non-linear approximation $\mathcal{A}'_N$ such that

$$\varepsilon'(2^p) \leq C 2^{-\alpha p}. \tag{5.46}$$

For $N = 2^p$, we define the set $E'_N(s)$ by

$$E'_N = \{0, 1, \ldots, 2^{p-1} - 1\} \cup E''_N(s) \tag{5.47}$$

where $E''_N(s)$ is a set of cardinal $|E''_N| \leq 2^{p-1}$ that depends on the locations of the discontinuities $\{d_1, d_2, \ldots, d_L\}$ in $s(t)$. To define $E''_N(s)$, we recall from Section 5.4 that the support of $\psi_{j,k}$ is in the interval $[2^{-j}(k-c), 2^{-j}(k+c)]$, and we consider the subset of discontinuities $\{d_1, d_2, \ldots, d_{L(p)}\}$ where

$$L(p) = \max\left\{ m \in \{1, \ldots, L\} \, \Big| \, m \leq \frac{2^{p-1}}{2c\alpha p} \right\}. \tag{5.48}$$

We define E''_N as the set of indices n of all wavelets $\psi_{j,k} = e_n$ such that $p - 1 \leq j < (\alpha+1)p - 1$ and $d_i \in \mathrm{supp}(\psi_{j,k})$ for some i in the set $\{1, \ldots, L(p)\}$. From the definition of $L(p)$, it is clear that $|E''_N| \leq 2^{p-1}$.

We now evaluate $\varepsilon'(2^p)$. Two events will be considered: $L(p) = L$ or $L(p) < L$.

If we have the event $L(p) = L$ and if the index n of a wavelet $\psi_{j,k} = e_n$ is not in E'_N, then two situations are possible:

• $d_i \notin \mathrm{supp}(\psi_{j,k})$ for all $i \in \{1, \ldots, L(p)\}$. This means that the support of $\psi_{j,k}$ is fully contained between d_i and d_{i+1} for some $i \in \{1, \ldots, L\}$, and from Theorem 5.1 we obtain the estimate

$$E(|\langle s, \psi_{j,k}\rangle|^2) = E(|\langle s_i, \psi_{j,k}\rangle|^2) \leq C 2^{-(\alpha+1)j}. \tag{5.49}$$

• $d_i \in \mathrm{supp}(\psi_{j,k})$ for some $i \in \{1, \ldots, L(p)\}$. In this case, we use the Schwarz inequality to obtain the crude estimate

$$E(|\langle s, \psi_{j,k}\rangle|^2) \leq E\left(\int_{(k-c)2^{-j}}^{(k+c)2^{-j}} |s(t)|^2 \, dt\right) \leq 2c\, r(0) 2^{-j}. \tag{5.50}$$

In view of the definition of E''_N, $j \geq (\alpha+1)p - 1$ in this case.

By summing (5.49) on k and on $j \geq p - 1$ and (5.50) on k and $j \geq (\alpha+1)p - 1$, we get the desired estimate for the conditional expectation, namely,

$$e_p^1 = E(\|\mathcal{A}'_N s - s\|_2^2 \mid L = L(p)) \leq C 2^{-\alpha p} \tag{5.51}$$

In the event that $L(p) < L$, that is, $L > \frac{2^{p-1}}{2c\alpha p}$, a crude estimate will be sufficient because of the small probability of this event as p goes to $+\infty$. We simply use

$$e_p^2 = E(\|\mathcal{A}'_N s - s\|_2^2 \mid L > L(p)) \leq E(\|s\|_2^2 \mid L > L(p)) = r(0). \tag{5.52}$$

Combining these estimates, we obtain

$$\begin{aligned}\varepsilon'(2^p) &= P(L = L(p))e_p^1 + P(L > L(p))e_p^2 \\ &\leq C\left(2^{-\alpha p} + P(L > \frac{2^{p-1}}{2c\alpha p})\right) \\ &= C\left(2^{-\alpha p} + e^{-\mu} \sum_{l > \frac{2^{p-1}}{2c\alpha p}} \frac{\mu^l}{l!}\right).\end{aligned}$$

One easily checks that the second term decreases exponentially faster than the first one, so that we finally have (5.46). This concludes the proof of the theorem. □

For the trigonometric system, $e_n(x) = e^{i2\pi nx}$, $n \in \mathbb{Z}$, we will show that non-linear approximation does not perform substantially better than linear projection. In fact, and in contrast with the wavelet basis that gives a sparse decomposition of the process $s(t)$, the best N Fourier coefficients essentially coincide with the first N Fourier coefficients. This fact is specific to the kinds of processes that we are considering. Other types of signals, such as velocity fields in turbulent flows, may present a more lacunary structure in the Fourier domain.

Theorem 5.4 *If $r(0) > r(1)$, then the non-linear approximation error for the trigonometric system satisfies the inequality*

$$\varepsilon(N) \geq CN^{-1}. \tag{5.53}$$

Proof. Define

$$s_k = \langle s, e_k \rangle = \int_0^1 s(t) e^{-i2\pi kt} \, dt. \tag{5.54}$$

We will prove that there exist strictly positive constants K and D such that the event

$$|k| \geq K \rightarrow |s_k| \geq Dk^{-1} \tag{5.55}$$

has a probability $p > 0$. The estimate (5.53) then follows from this property. Indeed, if we have the event (5.55), then for any set F_N

of cardinal $N > 0$,
$$\sum_{k \notin F_N} |s_k|^2 \geq CN^{-1}. \tag{5.56}$$
Applying (5.6) to E_N, we see that
$$\varepsilon(N) \geq pCN^{-1}, \tag{5.57}$$
which is equivalent to (5.53), up to a change in the constant C.

To show that (5.55) occurs with strictly positive probability, we consider the event where there is no discontinuity, i.e., $L = 0$. This event has the probability $e^{-\mu} > 0$. In this case, we decompose $s(t)$ into
$$s(t) = a(t) + b(t), \tag{5.58}$$
where
$$\left.\begin{array}{l} a(t) = (1-t)s(0) + ts(1) \\ b(t) = s(t) - (1-t)s(0) - ts(1). \end{array}\right\} \tag{5.59}$$
We denote the Fourier coefficients of $a(t)$ and $b(t)$ by a_k and b_k, $k \in \mathbb{Z}$, so that $s_k = a_k + b_k$.

From the assumption that $r(0) > r(1)$, we get
$$E(|s(0) - s(1)|^2 \mid L = 0) = 2(r(0) - r(1)) > 0, \tag{5.60}$$
and thus the event
$$|s(0) - s(1)| \geq \sqrt{r(0) - r(1)} \tag{5.61}$$
occurs with probability $p' > 0$. It is clear that (5.61) implies
$$|a_k| \geq 2Dk^{-1} \tag{5.62}$$
for some $D > 0$ related to $\sqrt{r(0) - r(1)}$.

We now consider the coefficients b_k. From (5.59), it is clear that $b(0) = b(1)$. Since $s(t)$ is differentiable in the mean square sense (in the case $L = 0$), we have for all $k \in \mathbb{Z} \setminus \{0\}$,
$$\begin{aligned} b_k &= \int_0^1 b(t) e^{-i2\pi kt} \, dt \\ &= (-i2\pi k)^{-1} \int_0^1 b'(t) e^{-i2\pi kt} \, dt \\ &= (-i2\pi k)^{-1} \int_0^1 s'(t) e^{-i2\pi kt} \, dt. \end{aligned}$$
This leads to
$$E(|b_k|^2) = \frac{E(|\langle s', e_k \rangle|^2)}{4\pi^2 k^2}. \tag{5.63}$$

Since $E(\|s'\|_2^2) < +\infty$, it follows that

$$\sum_{k\neq 0} k^2 E(|b_k|^2) < +\infty, \qquad (5.64)$$

and this implies, by Tchebycheff's inequality, that

$$\sum_{k\neq 0} P(|b_k| > Dk^{-1}) < +\infty. \qquad (5.65)$$

From (5.65), we can apply the Borel–Cantelli theorem to conclude that, for any $\rho > 0$, there exists K_ρ such that the event

$$|k| > K_\rho \rightarrow |b_k| \leq Dk^{-1}, \qquad (5.66)$$

has probability $1 - \rho$. We choose $\rho = p'/2$ so that (5.62) and (5.66) occur simultaneously with a probability greater than $p'/2$. Consequently, (5.55) is satisfied with $K = K_{p'/2}$ and $p \geq e^{-\mu}p'/2 > 0$.
□

REMARKS
- To obtain the estimate (5.55), we have used the regularity of $r(t)$, and, indeed, it seems difficult to avoid the assumption that r is twice differentiable.
- In contrast, the assumption that $r(0) > r(1)$ is not strictly necessary. It allows us to consider only the event $L = 0$. If this assumption is removed, the process could be periodic in the case $L = 0$. In this case, one still obtains (5.55) by working on the event $L = 1$. We kept the assumption that $r(0) > r(1)$ since in many practical situations $r(t)$ reaches its maximum only at the origin.

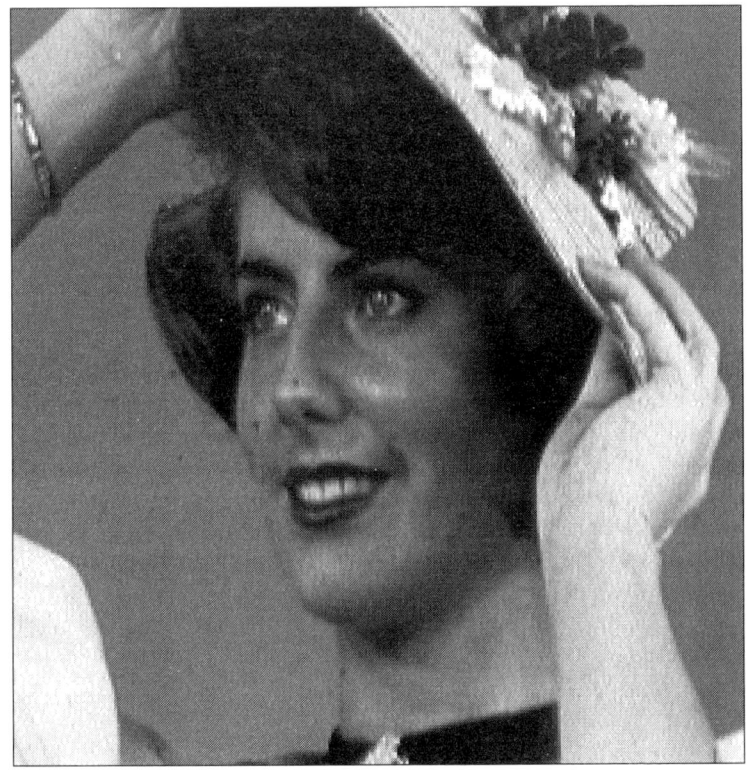

Figure 5.1 *Original image.*

NON-LINEAR APPROXIMATION

Figure 5.2 *Wavelet decomposition.*

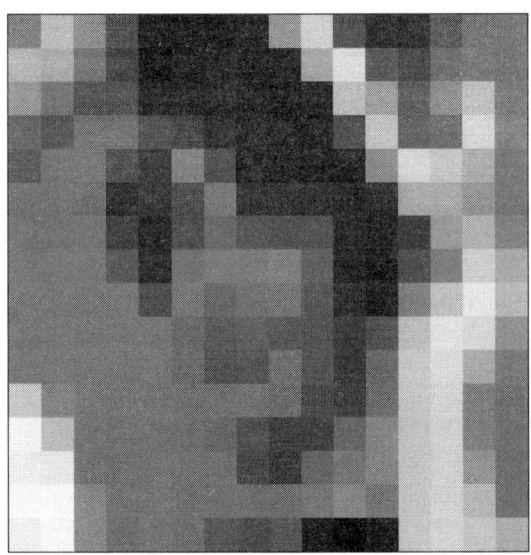

Figure 5.3a *Linear approximation with the Haar system, $j = 4$.*

Figure 5.3b *Linear approximation with the Haar system, $j = 5$.*

Figure 5.3c *Linear approximation with the Haar system, $j = 6$.*

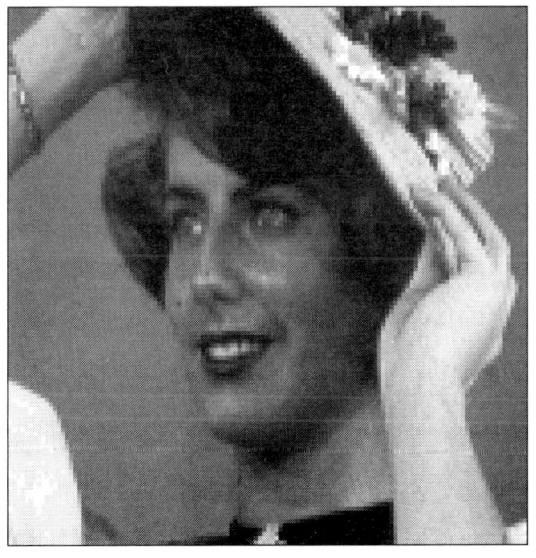

Figure 5.3d *Linear approximation with the Haar system, $j = 7$.*

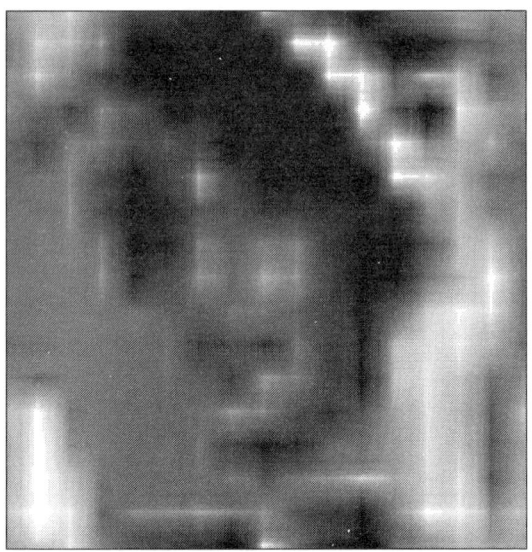

Figure 5.4a *Linear approximation with a regular wavelet basis, $j = 4$.*

Figure 5.4b *Linear approximation with a regular wavelet basis, $j = 5$.*

Figure 5.4c *Linear approximation with a regular wavelet basis*, $j = 6$.

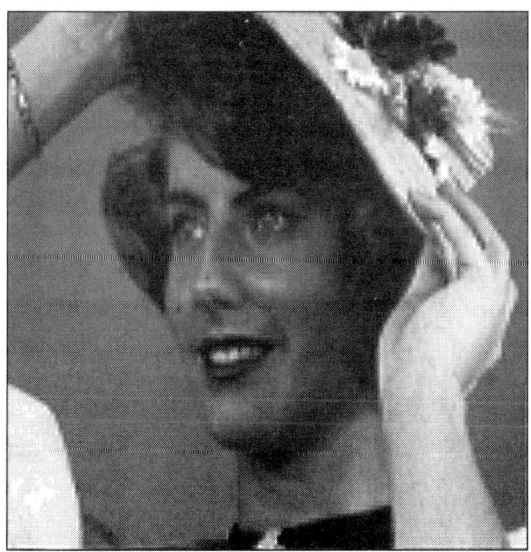

Figure 5.4d *Linear approximation with a regular wavelet basis*, $j = 7$.

Figure 5.5a *Linear approximation with a biorthogonal basis, $j = 4$.*

Figure 5.5b *Linear approximation with a biorthogonal basis, $j = 5$.*

NON-LINEAR APPROXIMATION

Figure 5.5c *Linear approximation with a biorthogonal basis*, $j = 6$.

Figure 5.5d *Linear approximation with a biorthogonal basis*, $j = 7$.

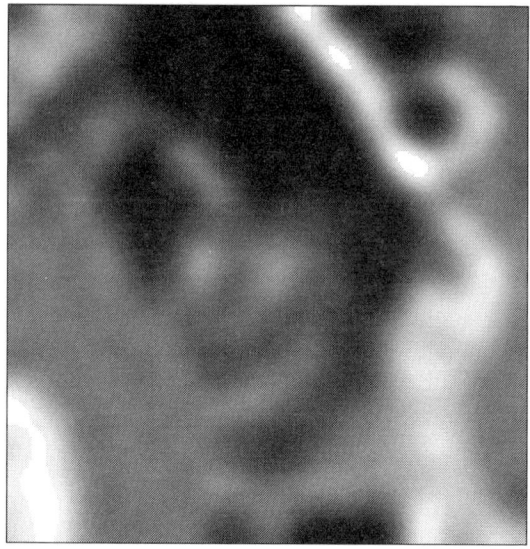

Figure 5.6a *Linear approximation with the trigonometric system, $j = 4$.*

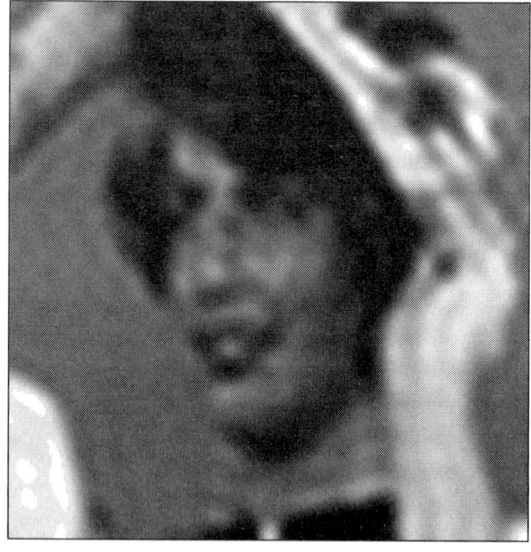

Figure 5.6b *Linear approximation with the trigonometric system, $j = 5$.*

NON-LINEAR APPROXIMATION

Figure 5.6c *Linear approximation with the trigonometric system,* $j = 6$.

Figure 5.6d *Linear approximation with the trigonometric system,* $j = 7$.

Figure 5.7 *Non-linear approximation with the Haar system.*

Figure 5.8 *Non-linear approximation with a regular wavelet basis.*

NON-LINEAR APPROXIMATION

Figure 5.9 *Non-linear approximation with a biorthogonal basis.*

Figure 5.10 *Non-linear approximation with the trigonometric system.*

APPENDIX A
Quasi-analytic wavelet bases

A complex signal f is said to be analytic, or progressive, if its Fourier transform satisfies $\hat{f}(\omega) = 0$ on the half-line $\omega \leq 0$. Analytic signals appear naturally in many fields of science and technology. Seismology and acoustics are examples. In these cases, one often wishes to analyze only the progressive (or forward moving) part of the signal since it corresponds to a certain physical reality.

The mathematical setting for studying these signals is the Hardy space $\mathcal{H}^2(\mathbb{R})$, which is defined by

$$\mathcal{H}^2(\mathbb{R}) = \left\{ f \in L^2(\mathbb{R}) \mid \hat{f}(\omega) = 0 \text{ for } \omega \leq 0 \right\}. \tag{A.1}$$

We note that if f is in this Hardy space, then

- f has an extension in the upperhalf-plane $\{\text{Im } z > 0\}$ as a holomorphic function; and
- $f(2^j x - k)$ is also in $\mathcal{H}^2(\mathbb{R})$ for all integers j and k.

This last remark might suggest the possibility of finding wavelet bases for the family of analytic signals. Unfortunately, one quickly sees that it is impossible to construct a localized multiresolution analysis, as presented in Chapter 1, for $\mathcal{H}^2(\mathbb{R})$: The scaling function φ must always satisfy the condition $|\hat{\varphi}(0)| = 1$, and thus $\hat{\varphi}$ would be discontinuous at the origin. More generally, it appears that there does not exist a wavelet basis $\{\psi_{j,k}\}$ for $\mathcal{H}^2(\mathbb{R})$ such that ψ is well localized in both time and frequency.

In the first applications of the continuous wavelet transform, Alex Grossmann and Jean Morlet (1984) used a Gaussian shifted in frequency that was 'practically analytic' in the sense that it was close enough to an analytic function to be suitable for numerical computations. On the other hand, this function could never generate an orthonormal basis.

We will show here that, if one is willing to allow an arbitrarily small part of the L^2 norm of the wavelet ψ to be located in the negative frequencies, then multiscale orthonormal bases for $L^2(\mathbb{R})$

can be constructed with functions that are both well localized in time and frequency.

More precisely, for each $\varepsilon > 0$, we will use Theorem 2.1 to construct a wavelet ψ_ε that belongs to the Schwartz class $\mathcal{S}(\mathbb{R})$ and that satisfies the relation

$$\omega \leq -\varepsilon \implies \hat{\psi}_\varepsilon(\omega) = 0. \tag{A.2}$$

Since $|\hat{\psi}(\omega)| \leq 1$, we see that $\int_{\omega \leq 0} |\hat{\psi}_\varepsilon(\omega)|^2\, d\omega \leq \varepsilon$ and that the same is true for the other functions in the orthonormal basis, namely

$$\int_{\omega \leq 0} |\widehat{(\psi_\varepsilon)}_{j,k}(\omega)|^2\, d\omega \leq \varepsilon. \tag{A.3}$$

This means that $L^2(\mathbb{R})$ can be generated with functions that each have an arbitrarily large spectral imbalance in favor of the positive frequencies. We suggest that the existence of such bases is not at all intuitive: The only previously known wavelets in the Schwartz class (the so-called 'Meyer wavelets') were generated by starting with their Fourier transforms that were assumed to have compact support and even modulus (see Meyer (1990)).

To construct ψ_ε, we will follow the standard procedure that uses a CQF defined by a regular function m_ε. We want the 2π-periodic function m_ε to satisfy the following property: For all ω in $[-\pi, \pi]$

$$m_\varepsilon(\omega) = 0 \iff \omega \in [-\pi, -\varepsilon/4]. \tag{A.4}$$

We assume that $0 < \varepsilon \leq 2\pi$ (see Figure A.1 at the end of this appendix).

Such a filter can be constructed with the help of a $\mathcal{C}^\infty$ function that 'links' 0 with 1 over the interval $[0, 1]$. For example, the function θ defined by

$$\theta(\omega) = \left[\int_0^1 \exp\left(\frac{-1}{\nu(1-\nu)}\right) d\nu\right]^{-1} \int_0^\omega \exp\left(\frac{-1}{\nu(1-\nu)}\right) d\nu \tag{A.5}$$

for $0 \leq \omega \leq 1$, $\theta(\omega) = 0$ for $\omega \leq 0$, and $\theta(\omega) = 1$ for $\omega \geq 1$.

Next, we define the function M_ε on $[-\pi, \pi]$ by

$$M_\varepsilon(\omega) = \theta\left(\frac{4\omega}{\varepsilon} + 1\right) \quad \text{for } -\pi \leq \omega \leq 0,$$

$$M_\varepsilon(\omega) = 1 - \theta\left(\frac{4(\omega - \pi)}{\varepsilon} + 1\right) \quad \text{for } 0 \leq \omega \leq \pi,$$

and we extend it to all of $\mathbb{R}$ by $M_\varepsilon(2\pi k + \omega) = M_\varepsilon(\omega)$. Then M_ε is a regular, 2π-periodic function that vanishes on the required set. Furthermore, from the properties of θ, it is clear that

$$M_\varepsilon(\omega) + M_\varepsilon(\pi + \omega) = 1 \tag{A.6}$$

so that

$$m_\varepsilon(\omega) = (M_\varepsilon(\omega))^{1/2} \tag{A.7}$$

looks like a possible choice for the CQF. Our first result shows that this filter produces the sought after basis.

Proposition A.1 *The CQF m_ε satisfies the hypotheses of Theorem 2.1, and the wavelet that it generates verifies property (A.2).*

Proof. Note that $m_\varepsilon(\omega) > 0$ for ω in the interval $\left[-\varepsilon/8, \pi - \varepsilon/8\right]$ (see Figure A.1). Thus we can choose

$$K_\varepsilon = \left[-\varepsilon/4, 2\pi - \varepsilon/4\right] \tag{A.8}$$

for the compact set congruent to $[-\pi, \pi]$ modulo 2π that one needs to conclude the results of Theorem 2.1, namely, that m_ε generates a localized multiresolution analysis. The associated scaling function φ_ε can be constructed from the formula

$$\hat{\varphi}_\varepsilon(\omega) = \prod_{k=1}^{+\infty} m_\varepsilon(2^{-k}\omega), \tag{A.9}$$

and $\hat{\varphi}_\varepsilon$ is in all of the Sobolev spaces $H^m(\mathbb{R})$. Since $\varepsilon \leq 2\pi$, $m_\varepsilon(\omega) = 0$ on the interval $[-\varepsilon/2, -\varepsilon/4] \subset [-\pi, -\varepsilon/4]$. This is enough to imply, by (A.9), that $\hat{\varphi}_\varepsilon(\omega) = 0$ for $\omega \leq -\varepsilon/2$.

The wavelet is defined by

$$\hat{\psi}_\varepsilon(\omega) = e^{-i\omega/2}\,\overline{m_\varepsilon(\omega/2 + \pi)}\,\hat{\varphi}_\varepsilon(\omega/2), \tag{A.10}$$

and thus it is clear that $\hat{\psi}_\varepsilon(\omega) = 0$ for $\omega \leq -\varepsilon$. □

Figure A.3, which represents $\hat{\psi}_\varepsilon(\omega)$, shows that the negative frequencies are eventually covered by dilations of $\hat{\psi}_\varepsilon$. However the high frequencies are only attained for very large values of j.

Proposition A.2 *The functions φ_ε and ψ_ε belong to the Schwartz class $\mathcal{S}(\mathbb{R})$.*

Proof. We will prove this result for φ_ε; the result for ψ_ε follows from (A.10).

Notice first that $\hat{\varphi}_\varepsilon$ vanishes on the intervals $[2^n\pi, 2^{n+1}\pi - \varepsilon/2]$ for all $n \in \mathbb{N}^*$. This follows by induction. Indeed, the function

$m_\varepsilon(\omega/2)$ is zero on the interval $[2\pi, 4\pi - \varepsilon/2]$, and thus by (A.9), so is $\hat\varphi_\varepsilon(\omega)$.

Assume that $\hat\varphi_\varepsilon(\omega) = 0$ when ω is in $[2^{n-1}\pi, 2^n\pi - \varepsilon/2]$ where $n \geq 2$. Then $\hat\varphi_\varepsilon(\omega/2) = 0$ on $[2^n\pi, 2^{n+1}\pi - \varepsilon]$. We know by construction that $m_\varepsilon(\omega/2) = 0$ on $[2^{n+1}\pi - 2\pi, 2^{n+1}\pi - \varepsilon/2]$ for all n. Consequently, $\hat\varphi_\varepsilon(\omega) = \hat\varphi_\varepsilon(\omega/2)\, m_\varepsilon(\omega/2)$ vanishes on the interval $[2^n\pi, 2^{n+1}\pi - \varepsilon/2]$. Note that here we use the assumption that ε is less than or equal to 2π.

We have shown that the only places where $\hat\varphi_\varepsilon$ does not vanish (other than the interval at the origin) are the points $\omega_n = 2^n\pi - \omega$ for $0 < \omega < \varepsilon/2$. We now examine the decay of $\hat\varphi_\varepsilon$ and its successive derivatives at these points. To do this, we use the following two estimates.

For each m in $\mathbb{N}$ there exists a constant C_m such that

$$\left|\left(\frac{d}{d\omega}\right)^m(\hat\varphi_\varepsilon)(\omega)\right| \leq C_m \sum_{k=0}^m \left|\left(\frac{d}{d\omega}\right)^k(m_\varepsilon)(2^{-n}\omega)\right| \quad (A.11)$$

for all $\omega > 0$ and for all n in $\mathbb{N}^*$, and for each m and k in $\mathbb{N}$ there exists a constant $C_{m,k}$ such that

$$\left|\left(\frac{d}{d\omega}\right)^m(m_\varepsilon)(\pi - \omega)\right| \leq C_{m,k}|\omega|^k \quad (A.12)$$

for all ω. (We will show how these estimates are derived in a moment.) Assuming that (A.11) and (A.12) are established, apply (A.11) and then (A.12) to the points $\omega_n = 2^n\pi - \omega$. Then we see that

$$\left|\left(\frac{d}{d\omega}\right)^m(\hat\varphi_\varepsilon)(\omega_n)\right| \leq B_{m,k}\, 2^{-nk}|\omega|^k \leq D_{m,k}|\omega_n|^{-k}, \quad (A.13)$$

where $B_{m,k} = C_m \sum_{l=0}^m C_{l,k}$ and $D_{m,k}$ can be taken to be $\pi^{2k} B_{m,k}$. This last inequality means that $\hat\varphi_\varepsilon$ belongs to the Schwartz class $\mathcal{S}(\mathbb{R})$. Since $\mathcal{S}(\mathbb{R})$ is invariant under the Fourier transform, we conclude that φ_ε and ψ_ε are also elements of $\mathcal{S}(\mathbb{R})$.

To finish the proof, we need to indicate how the two estimates (A.11) and (A.12) are obtained. The second one, (A.12), is easily derived from the definition of θ and hence of m_ε. The point is that the zero of θ at 0 (and hence of m_ε at π) is of infinite order. More precisely, the derivative of θ at zero looks like $\exp(-\frac{1}{\omega})$, meaning that $\theta(\omega)$ goes to zero as ω tends to zero faster than any power of ω. This translates directly into the inequality (A.12).

The other estimate, (A.11), is conceptually simple but tedious

QUASI-ANALYTIC WAVELET BASES

to derive completely because of the heavy notation. We will do the case $m = 1$ to give an idea of how things go.

The first thing to observe is that the infinite product

$$\hat{\varphi}_\varepsilon(\omega) = \prod_{k=1}^{+\infty} m_\varepsilon(2^{-k}\omega) \qquad (A.14)$$

is in fact finite for any given $\omega > 0$ and for any compact neighborhood of $\omega > 0$. This is because, for a fixed ω, $m_\varepsilon(2^{-k}\omega) = 1$ for all sufficiently large k. With this in mind, we need not be concerned about convergence of the following sums and products, and although we write these as being infinite, they are in fact finite. Thus we can write

$$\frac{d\hat{\varphi}_\varepsilon}{d\omega}(\omega) = \sum_{k=1}^{\infty} 2^{-k} \frac{dm_\varepsilon}{d\omega}(2^{-k}\omega) \prod_{\substack{l=1 \\ l \neq k}}^{+\infty} m_\varepsilon(2^{-l}\omega). \qquad (A.15)$$

When we display the terms on the right-hand side whose argument is $2^{-n}\omega$ we get

$$\frac{d\hat{\varphi}_\varepsilon}{d\omega}(\omega) = m_\varepsilon(2^{-n}\omega)\left[\sum_{\substack{k=1 \\ k \neq n}}^{\infty} 2^{-k} \frac{dm_\varepsilon}{d\omega}(2^{-k}\omega) \prod_{\substack{l=1 \\ l \neq k,n}}^{+\infty} m_\varepsilon(2^{-l}\omega)\right]$$

$$+ \frac{dm_\varepsilon}{d\omega}(2^{-n}\omega)2^{-n} \prod_{\substack{l=1 \\ l \neq n}}^{+\infty} m_\varepsilon(2^{-l}\omega).$$

If we let $M^{(1)} = \sup_{\omega \in \mathbb{R}} |\frac{dm_\varepsilon}{d\omega}(\omega)|$ and use the fact that $|m_\varepsilon(\omega)| \leq 1$, then we see that

$$\left|\frac{d\hat{\varphi}_\varepsilon}{d\omega}(\omega)\right| \leq M^{(1)}|m_\varepsilon(2^{-n}\omega)| + 2^{-n}\left|\frac{dm_\varepsilon}{d\omega}(2^{-n}\omega)\right|$$

$$\leq \max\{M^{(1)}, 1/2\} \sum_{k=0}^{1} \left|\left(\frac{d}{d\omega}\right)^k (m_\varepsilon)(2^{-n}\omega)\right|.$$

This is the result for $m = 1$. The general result involves the same ingredients, namely Leibniz's formula for computing the derivatives of products, the fact that $|m_\varepsilon(\omega)| \leq 1$, the fact that the derivatives of m_ε are bounded, and the fact that each differentiation of $m_\varepsilon(2^{-k}\omega)$ introduces the factor 2^{-k}. In general, the constants C_m will be functions of the numbers $M^{(k)} = \sup_{\omega \in \mathbb{R}} |(\frac{d}{d\omega})^m (m_\varepsilon)(\omega)|$ for $k = 1, 2, \ldots, m$. □

REMARKS
- While almost all of the energy of the wavelet can be concentrated in the positive half of the spectral domain, this is not true for the function $\frac{|\hat{\psi}(\omega)|^2}{|\omega|}$, whose integral is always divided equally between the left and right halves of the axis.

This is in fact a necessary condition so that the dilations of $\hat{\psi}$ cover the whole frequency axis. This is seen from the following general argument.

The fundamental equation $|m_0(\omega)|^2 + |m_0(\omega + \pi)|^2 = 1$ implies, on multiplying both sides by $|\hat{\varphi}(\omega)|^2$, that

$$|\hat{\varphi}(\omega)|^2 = |\hat{\psi}(2\omega)|^2 + |\hat{\varphi}(2\omega)|^2, \qquad (A.16)$$

which, in turn, implies that

$$|\hat{\varphi}(2^{k-1}\omega)|^2 - |\hat{\varphi}(2^k\omega)|^2 = |\hat{\psi}(2^k\omega)|^2 \qquad (A.17)$$

for all $k \in \mathbb{Z}$. Summing both sides and using the facts that $\hat{\varphi}(2^k\omega)$ tends to 1 when k tends to $-\infty$ and that (for $\omega \neq 0$) $\hat{\varphi}(2^k\omega)$ tends to 0 when k tends to $+\infty$, we obtain the identity

$$\sum_{j \in \mathbb{Z}} |\hat{\psi}(2^j \omega)|^2 = 1 \qquad (A.18)$$

for all $\omega \neq 0$. This identity allows us to compute explicitly the integrals of $\frac{|\hat{\psi}(\omega)|^2}{|\omega|}$ on the positive and negative axes. Thus,

$$\begin{aligned}
\int_{\omega \geq 0} \frac{|\hat{\psi}(\omega)|^2}{|\omega|} d\omega &= \sum_{j \in \mathbb{Z}} \int_{2^j}^{2^{j+1}} \frac{|\hat{\psi}(\omega)|^2}{|\omega|} d\omega \\
&= \int_1^2 \frac{\sum_{j \in \mathbb{Z}} |\hat{\psi}(2^j\omega)|^2}{|\omega|} d\omega \\
&= \log 2.
\end{aligned}$$

The same computation shows that $\int_{\omega \leq 0} \frac{|\hat{\psi}(\omega)|^2}{|\omega|} d\omega = \log 2$.

The use of the measure $\frac{d\omega}{|\omega|}$, which is invariant under dilations, thus leads to an equal 'energy' distribution between the positive and negative frequencies.

- When the parameter ε becomes very small, the function m_ε approaches a step function, which corresponds to a very poor resolution in the time domain. One can thus expect to encounter numerical problems when using these filters even though the transfer function is formally regular.

QUASI-ANALYTIC WAVELET BASES 205

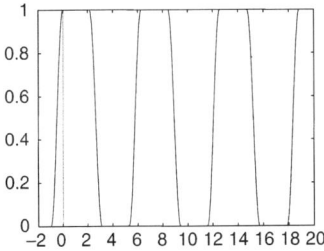

Figure A.1 $m_\varepsilon(\omega)$.

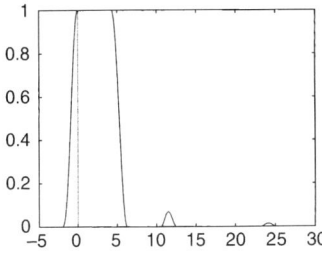

Figure A.2 $\hat{\varphi}_\varepsilon(\omega)$.

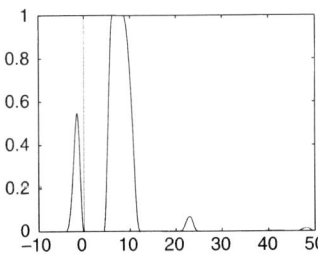

Figure A.3 $\hat{\psi}_\varepsilon(\omega)$.

APPENDIX B

Multivariate constructions

We present several examples of multivariate wavelets that are nontrivial in the sense that they are not tensor products of wavelets constructed with one variable.

Recall from Chapter 1 that if we wish to generalize the CQFs for n variables, the defining equation becomes

$$\sum_{(\varepsilon_1,\ldots,\varepsilon_n)\in\{0,1\}^n} |m_0(\omega_1+\varepsilon_1\pi,\ldots,\omega_n+\varepsilon_n\pi)|^2 = 1. \quad (\text{B.1})$$

The complete classification of the finite impulse response filters that satisfy this property remains an open, and seemingly difficult, question. In the univariate case, this classification is based on the Riesz lemma that ensures that any positive polynomial in $\cos\omega$ is the squared modulus of a trigonometric polynomial $p_0(e^{i\omega})$. Unfortunately, no similar result holds in the multivariate setting. Hence, one cannot hope to derive much from an explicit characterization of $|m_0(\omega)|^2$, which itself is not obvious.

We will see, however, that one can significantly extend the class of multiresolution analyses obtained with tensor products by using well-chosen transformations. These constructions are based on Proposition B.1.

Proposition B.1 *Let $\mathcal{V}_j = V_j^1 \otimes \cdots \otimes V_j^n$ be a multiresolution of $L^2(\mathbb{R}^n)$ obtained from the localized multiresolution analyses $\{V_j^i\}$ of $L^2(\mathbb{R})$, $i = 1,\ldots,n$.*

Denote the associated scaling function by

$$\phi(x) = \phi(x_1,\ldots,x_n) = \varphi^1(x_1)\cdots\varphi^n(x_n). \quad (\text{B.2})$$

Let $T = (t_{ij})_{i,j=1,\ldots,n}$ be an $n\times n$ matrix with integer coefficients, and define the scaling function ϕ_T by

$$\hat{\phi}_T(\omega) = \hat{\phi}(T\omega). \quad (\text{B.3})$$

Then the associated sequence $\mathcal{V}_j^T$ is a multiresolution analysis if and only if $|\det(T)| = 1$.

Proof. Clearly the determinant of T cannot be zero: In this case one would have $\hat{\phi}_T(\omega + \omega_0) = \hat{\phi}_T(\omega)$ for all ω_0 in Ker(T), and the function ϕ_T could not belong to $L^2(\mathbb{R}^n)$. Once this case is eliminated, it is clear that ϕ_T satisfies the localization condition ($\hat{\phi}_T$ is in all of the Sobolev spaces) and that $\left| \int_{\mathbb{R}^n} \phi_T(y)\, dy \right| = 1$. The latter can be seen by noting that $\phi_T(x) = \phi(S^t x) |\det S|$ and letting $y = S^t x$, where $S = T^{-1}$ and S^t is the transpose of S.

The problem then reduces by Theorem 1.1 (which is easily generalized) to the question of the orthogonality of ϕ_T and its translates.

- If $|\det(T)| = 1$, then

$$\sum_{k \in \mathbb{Z}^n} |\hat{\phi}_T(\omega + 2k\pi)|^2 = \sum_{k \in \mathbb{Z}^n} |\hat{\phi}(T\omega + 2\pi T k)|^2$$
$$= \sum_{k \in \mathbb{Z}^n} |\hat{\phi}(T\omega + 2k\pi)|^2$$
$$= 1$$

because T is a bijection of $\mathbb{Z}^n$ onto itself. But this means that ϕ_T and its translates are orthogonal and, hence, that ϕ generates a multiresolution analysis.

- If $|\det(T)| > 1$, the norm of the functions $\phi_T(x - k)$ is not 1: Using the notation introduced above, we see that

$$\int |\phi_T(x)|^2\, dx = \int |\phi(S^t x)|^2 |\det S|^2\, dx$$
$$= \frac{1}{|\det(T)|} \int |\phi(y)|^2\, dy$$
$$= \frac{1}{|\det(T)|}\,.$$

Furthermore, when $|\det(T)| > 1$, T is not a bijection of $\mathbb{Z}^n$ onto itself, and one can thus choose an $\alpha \in \mathbb{R}^n$ such that $T\alpha \in 2\pi \mathbb{Z}^n$ and $T\alpha \notin T(2\pi \mathbb{Z}^n)$. Then we have

$$\sum_{k \in \mathbb{Z}^n} |\hat{\phi}_T(\alpha + 2k\pi)|^2 = 0\,, \tag{B.4}$$

which means that $\{\phi_T(x - k)\}_{k \in \mathbb{Z}}$ is not even a Riesz basis.

Proposition B.1 allows one to create a more general class of wavelets that go beyond the obvious tensor products. The associated filters are, of course, obtained by using the same transformations. One thing that can be done by an appropriate choice of T is to change the spectral orientation of a filter. For example, we have seen in Proposition 3.6 that the CQFs introduced by Ingrid Daubechies approach, as their size increases, a perfect band-pass filter that eliminates the high frequencies and conserves the spectral energy in the band $\left[-\frac{\pi}{2}, \frac{\pi}{2}\right]$. Their tensor product thus conserves, for the most part, the frequencies in the hypercube $\left[-\frac{\pi}{2}, \frac{\pi}{2}\right]^n$. Under the action of T, we obtain a new filter that passes the frequencies in $T^{-1}\left(\left[-\frac{\pi}{2}, \frac{\pi}{2}\right]^n\right)$. By an astute choice for T^{-1}, we can favor a given direction in frequency space.

We will use the bivariate case to illustrate this assertion. Thus, consider a prescribed direction represented by a vector (x, y) in $\mathbb{R}^2 \setminus \{0\}$. This direction can be approached arbitrarily closely by a vector (p, q) where p and q are relatively prime integers.

By Bezout's theorem, there exist two other integers r and s such that $ps - qr = 1$. Thus we can take T^{-1} to be

$$T^{-1} = \begin{pmatrix} p & r \\ q & s \end{pmatrix}. \tag{B.5}$$

By taking $\sup(|p|, |q|)$ large enough, (p, q) will approximate the direction (x, y) as closely as desired. Furthermore, the parallelogram $T^{-1}\left(\left[-\frac{\pi}{2}, \frac{\pi}{2}\right]^2\right)$ is concentrated around the direction (p, q), while its area remains equal to π^2.

This construction can be generalized to n variables. We simply notice that an arbitrary direction $(x_1, \ldots, x_n)$ can be approached as closely as we wish by a vector $(p_1, \ldots, p_n)$ where each pair of integers p_i and p_j, $i \neq j$, are relatively prime and $\sup(|p_i|)$ is sufficiently large. We can then find a matrix with integer coefficients and determinant equal to 1 that has this vector as its first column. For example, we simply take

$$T^{-1} = \begin{pmatrix} p_1 & q_1 & 0 & \cdots & & \cdots & 0 \\ p_2 & q_2 & 0 & & & & \vdots \\ p_3 & 0 & 1 & \ddots & & & \vdots \\ \vdots & \vdots & \ddots & \ddots & \ddots & & \vdots \\ \vdots & \vdots & & & \ddots & 1 & 0 \\ p_n & 0 & \cdots & & \cdots & 0 & 1 \end{pmatrix} \tag{B.6}$$

and choose q_1 and q_2 so that $p_1 q_2 - p_2 q_1 = 1$.

This result is in the same spirit as the construction of quasi-analytic wavelet bases in Appendix A, in the sense that it shows that we can bias the analysis is certain directions. But contrary to the quasi-analytic construction, here we can use finite impulse response filters.

We show an example of a possible transformation for the bivariate case in Figure B.1.

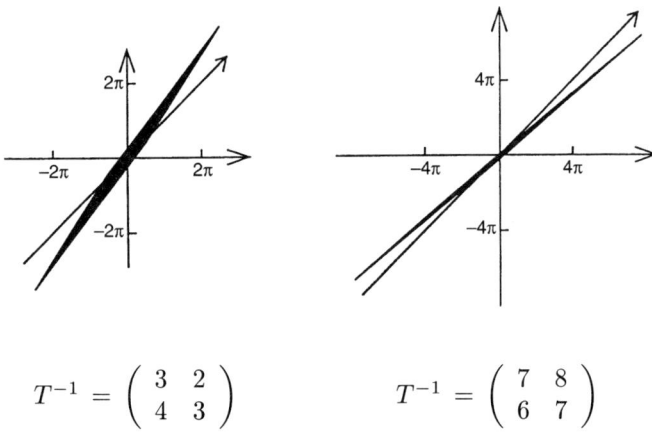

$$T^{-1} = \begin{pmatrix} 3 & 2 \\ 4 & 3 \end{pmatrix} \qquad T^{-1} = \begin{pmatrix} 7 & 8 \\ 6 & 7 \end{pmatrix}$$

Figure B.1 *Spectral concentration around the direction* $(1,1)$.

We note that for the biorthogonal case, which we developed in Chapter 4, the problem of constructing non-separable filters is simpler to solve than in the orthogonal case. Recall that solving for the coefficients is a linear problem and that the Riesz lemma is not needed.

More precisely, a multivariate signal is filtered using 2^n digital filters $\{H_i(\omega)\}_{i=0,\ldots,2^n-1}$ and the result is then decimated in the ratio of 1 to 2^n. The condition for perfect reconstruction using filters with finite impulse response is expressed by

$$\det(H_i(\omega + \varepsilon_j \pi))_{\substack{i=0,\ldots,2^n-1 \\ \varepsilon_j \in \{0,1\}^n}}$$
$$= \begin{vmatrix} H_0(\omega_1,\ldots,\omega_n) & \cdots & H_0(\omega_1+\pi,\ldots,\omega_n+\pi) \\ \vdots & & \vdots \\ H_{2^n-1}(\omega_1,\ldots,\omega_n) & \cdots & H_{2^n-1}(\omega_1+\pi,\ldots,\omega_n+\pi) \end{vmatrix}$$
$$= 1.$$

If the first $2^n - 1$ filters are fixed, solving for the last one can be considered a Bezout problem, and this has a solution if the $2^n - 1$ minors obtained from the $2^n - 1$ first rows are relatively prime polynomials. The coefficients of H_{2^n-1} are then obtained by solving a linear problem.

We next present a completely different approach for extending the possible multiresolution analyses on $L^2(\mathbb{R}^n)$. This consists in using dilation matrices different from $2I_n$. One considers a matrix $D = (m_{ij})_{i,j=1,\ldots,n}$ with integer coefficients and whose eigenvalues have absolute value strictly greater than 1.

It is then possible to define a multiresolution analysis with respect to D: One keeps all the axioms of Chapter 1 except for the second one, which is replaced by

$$f(x) \in V_j \iff f(Dx) \in V_{j+1}. \tag{B.7}$$

Starting with these new axioms, one can build the scaling function φ, the wavelets, and the associated filters. One can also establish results analogous to those in Chapters 2, 3, and 4. However, many new difficulties that are specific to the multivariate setting arise, particularly when estimating the regularity of the scaling functions (see Cohen and Daubechies (1993)).

An interesting aspect of this generalization comes from the fact that the number of wavelets that are needed to generate the complement of V_j in V_{j+1} equals $\det(D) - 1$. This is because the decimation in going from one scale to the next larger one is in the ratio 1 to $\det(D)$. Thus, by choosing a matrix with determinant 2, one can hope to generate $L^2(\mathbb{R}^n)$ starting with a single function and constructing the basis $\{|\det(D)|^{j/2} \psi(D^j x - k)\}_{j \in \mathbb{Z}, k \in \mathbb{Z}^n}$.

Here we focus on the case of orthonormal wavelets and scaling functions. A simple example of such a construction is given by the matrix

$$D = \begin{pmatrix} 0 & 1 & 0 & \ldots & 0 \\ \vdots & 0 & \ddots & \ddots & \vdots \\ \vdots & \vdots & \ddots & \ddots & 0 \\ 0 & 0 & \ldots & 0 & 1 \\ 2 & 0 & \ldots & \ldots & 0 \end{pmatrix} = \begin{pmatrix} 0 & I_{n-1} \\ 2 & 0 \end{pmatrix}, \tag{B.8}$$

which satisfies $D^n = 2I_n$. The action of this matrix dilates one direction and permutes the coordinates. This choice brings us directly back to the univariate case because the process effects a dilation in each direction separately. The filter one uses is simply $M_0(\omega_1, \ldots, \omega_n) = m_0(\omega_1)$ where m_0 is a univariate CQF. The scaling function is given by

$$\hat{\phi}(\omega_1, \ldots, \omega_n) = \prod_{j=1}^{+\infty} M_0(D^{-j}(\omega_1, 0, \ldots, 0))$$

$$= \prod_{j=1}^{+\infty} \prod_{k=1}^{n} m_0(2^{-j}\omega_k)$$

$$= \hat{\varphi}(\omega_1) \cdots \hat{\varphi}(\omega_n) .$$

One can look for less trivial constructions by using matrices that are more 'isotropic.' An example in two variables is given by the matrix $D = \begin{pmatrix} 1 & -1 \\ 1 & 1 \end{pmatrix}$ that was mentioned in Section 1.4. In this case, the equation that defines the filter m_0 is

$$|M_0(\omega_1, \omega_2)|^2 + |M_0(\omega_1 + \pi, \omega_2 + \pi)|^2 = 1. \quad (B.9)$$

Here one is tempted to choose for $M_0(\omega_1, \omega_2)$ a univariate filter $m_0(\omega_1)$ that satisfies (B.9) trivially. If one does this by choosing one of the filters derived by Ingrid Daubechies (1988, 1992), a problem concerning the orthonormality of $\{\varphi(x-k)\}_{k \in \mathbb{Z}^2}$ appears.

Recall that the family of CFQs constructed by Ingrid Daubechies is given by

$$|m_{0,N}(\omega)|^2 = \left(\cos^2 \frac{\omega}{2}\right)^N \sum_{j=0}^{N-1} \binom{N-1+j}{j} \left(\sin^2\left(\frac{\omega}{2}\right)\right)^j .$$
(B.10)

The functions $m_{0,N}$ vanish only at the points $-\pi$ and π, and the hypotheses of Theorem 2.1 are trivially satisfied in the univariate case by taking $[-\pi, \pi]$ as the compact set K congruent to $[-\pi, \pi]$ modulo 2π.

In the case of two variables, condition (P) of Theorem 2.1 is easily reformulated:

> There exists a compact set K congruent to $[-\pi, \pi]^2$ modulo 2π whose interior contains 0 and such that, and for all $\omega = (\omega_1, \omega_2)$ in K, $M_0(D^{-j}(\omega_1, \omega_2)) \neq 0$ for all $j \geq 1$. (P′)

This time the choice $K = [-\pi, \pi]^2$ does not work, since clearly

$$M_0(D^{-1}(\pi, \pi)) = M_0(D^{-1}(-\pi, -\pi)) = m_0(\pi) = 0. \quad (B.11)$$

Thus, to satisfy condition (P'), it is necessary to construct a compact set K that is a bit more complicated. We do this by translating small triangular neighborhoods of the points $(-\pi, -\pi)$ and (π, π), which are the only points presenting a problem. This construction is shown in Figure B.2. Once this is done, Theorem 2.1 guarantees the existence of orthonormal bases.

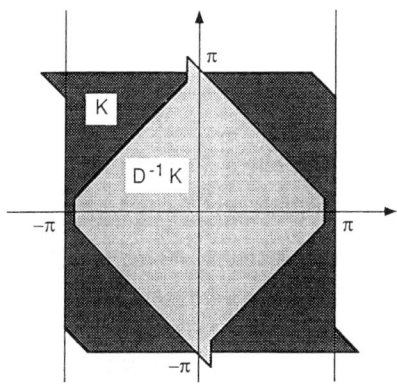

Figure B.2 *Construction of the compact set K.*

As a final note we mention a result discovered by K. Gröchenig and W. R. Madych (1992): If M_0 is taken to be the simplest possible filter, that which corresponds to the univariate Haar system, then the scaling function φ is the characteristic function of a set E whose frontier is a fractal curve.

The defining equation for φ is

$$\varphi(X) = \varphi(DX) + \varphi(DX + (1,0)), \tag{B.12}$$

and it is possible to verify directly that the set E is described in the complex plane by

$$E = \left\{ \sum_{k=1}^{+\infty} \varepsilon_k \left(\frac{1+i}{2}\right)^k \; \middle| \; \varepsilon_k = 0 \text{ or } 1 \right\}. \tag{B.13}$$

Figure B.3 represents the set E. It is equal to the juxtaposition of its two contractions $D^{-1}E$ and $D^{-1}(E-(1,0))$. This self-similar structure is expressed by the equation (B.12). What is more, the integer translates of E provide a perfect tiling of the plane since $|E| = 1$ and $\langle \mathbf{1}_E, \mathbf{1}_{E+k} \rangle = \delta_{0,k}$.

More generally, such tilings can be obtained as solutions of equations
$$\varphi(X) = \sum_{\gamma_i \in \Gamma} \varphi(DX - \gamma_i)$$
where D is a dilation matrix with integer entries and Γ is a set of representatives of $\mathbb{Z}^n \setminus D\mathbb{Z}^n$. The choice of Γ must be made so that the condition (P') is satisfied.

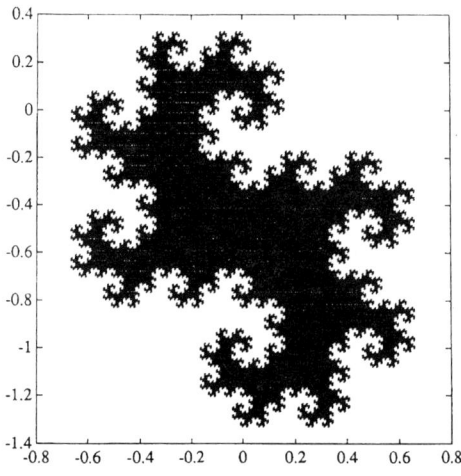

Figure B.3 *The fractal set E.*

APPENDIX C

Multiscale unconditional bases

The purpose of this appendix is to present a mathematical application of the biorthogonal systems introduced in Chapter 4. Specifically, we will construct unconditional bases for the Sobolev space $H^1(\mathbb{R})$ that have a multiscale structure $\{\psi_{j,k}\}_{j,k\in\mathbb{Z}}$ and where the function ψ is as simple as possible.

It is clear that the Haar system is not suitable since these functions do not belong to $H^1(\mathbb{R})$. We will thus use a slightly more complex function, one that is piecewise affine and that has enough oscillation so that the $\psi_{j,k}$ are 'asymptotically orthogonal' as the ratio of the scales tends to infinity.

We denote the triangle function by $\Delta(x) = (1 - |x|)\,\mathbf{1}_{[-1,1]}$. The two simplest choices for ψ are then

$$\psi_a(x) = \Delta(2x) - \Delta(2x-1), \qquad (C.1)$$

if the degree of cancellation for ψ is fixed at 1, and

$$\psi_b(x) = \Delta(2x-1) - \frac{1}{2}\Delta(2x) - \frac{1}{2}\Delta(2x-2), \qquad (C.2)$$

if a second vanishing moment is required. (See the figures at the end of this appendix.)

We will use the theory of biorthogonal wavelets to show that the second function generates an unconditional multiscale basis for $H^1(\mathbb{R})$ and that the dual basis (with respect to the scalar product in $L^2(\mathbb{R})$) has the same structure.

Recall that, in the constructions of Chapter 4, the definition of the function ψ involves both the filter m_0 and its dual $\tilde{m}_0$:

$$\hat{\psi}(2\omega) = e^{-i\omega}\,\overline{\tilde{m}_0(\omega+\pi)}\,\prod_{k=1}^{+\infty} m_0(2^{-k}\omega). \qquad (C.3)$$

If we choose $m_0(\omega) = \left|\frac{1+e^{i\omega}}{2}\right|^2$ so that the infinite product generates the triangle function, we see that the choices (C.1) and (C.2) for ψ determine the dual filters $\tilde{m}_0$. Unfortunately, neither of these

filters satisfy the essential duality relation
$$m_0(\omega)\overline{\tilde{m}_0(\omega)} + m_0(\omega+\pi)\overline{\tilde{m}_0(\omega+\pi)} = 1. \qquad (C.4)$$
Consequently, we cannot conclude directly from Theorem 4.1 that the functions ψ_a and ψ_b generate unconditional bases.

Fortunately, there is a simple result that allows us to significantly enlarge the set of biorthogonal, multiscale unconditional bases.

Proposition C.1 *Assume that $\{\psi_{j,k}, \tilde{\psi}_{j,k}\}_{j,k \in \mathbb{Z}}$ is a biorthogonal system whose scaling functions satisfy the decay hypotheses of Theorem 4.1 and let m be a 2π-periodic, $\mathcal{C}^\infty(\mathbb{R})$ function that does not vanish.*

Define the functions ψ_m and $\tilde{\psi}_m$ by
$$\hat{\psi}_m(\omega) = m(\omega)\hat{\psi}(\omega), \qquad (C.5)$$
and
$$\widehat{\tilde{\psi}_m}(\omega) = (\overline{m(\omega)})^{-1}\widehat{\tilde{\psi}}(\omega). \qquad (C.6)$$
Then the system $\{(\psi_m)_{j,k}, (\tilde{\psi}_m)_{j,k}\}_{j,k \in \mathbb{Z}}$ is a biorthogonal unconditional basis.

Proof. The first thing to observe — and this is easily seen by taking Fourier transforms — is that $\{(\psi_m)_{j,k}\}_{k \in \mathbb{Z}}$ and $\{\psi_{j,k}\}_{k \in \mathbb{Z}}$ span the same subspace, W_j, for each fixed $j \in \mathbb{Z}$. This relies on m being a smooth, 2π-periodic function that is bounded away from zero. Next, since $\{\psi_{j,k}\}_{k \in \mathbb{Z}}$ is by assumption an unconditional basis for W_j, there exist strictly positive constants K_1 and K_2 such that
$$K_1 \sum_{k \in \mathbb{Z}} |\alpha_k|^2 \leq \left\|\sum_{k \in \mathbb{Z}} \alpha_k \psi_{j,k}(x)\right\|^2 \leq K_2 \sum_{k \in \mathbb{Z}} |\alpha_k|^2.$$
The constants K_1 and K_2 do not depend on j since
$$\left\|\sum_{k \in \mathbb{Z}} \alpha_k \psi_{j,k}(x)\right\|^2 = \left\|\sum_{k \in \mathbb{Z}} \alpha_k \psi_k(x)\right\|^2.$$
These last two relations, combined with the equation
$$\left\|\sum_{k \in \mathbb{Z}} \alpha_k (\psi_m)_{j,k}(x)\right\|^2 = \frac{1}{4\pi^2} \int \left|\sum_{k \in \mathbb{Z}} \alpha_k e^{-ik\omega}\right|^2 |m(\omega)|^2 |\hat{\psi}(\omega)|^2 d\omega,$$
show that
$$c^2 K_1 \sum_{k \in \mathbb{Z}} |\alpha_k|^2 \leq \left\|\sum_{k \in \mathbb{Z}} \alpha_k (\psi_m)_{j,k}(x)\right\|^2 \leq C^2 K_2 \sum_{k \in \mathbb{Z}} |\alpha_k|^2,$$

where $c = \min |m(\omega)| \neq 0$ and $C = \max |m(\omega)|$. This proves that $\{(\psi_m)_{j,k}\}_{k\in\mathbb{Z}}$ is an unconditional basis for W_j. A parallel argument with c replaced by $1/C$ and C replaced by $1/c$ proves that $\{(\tilde{\psi}_m)_{j,k}\}_{k\in\mathbb{Z}}$ is an unconditional basis for $\tilde{W}_j$.

We thus have the equivalences

$$C_1 \sum_{k\in\mathbb{Z}} |\alpha_k|^2 \leq \left\|\sum_{k\in\mathbb{Z}} \alpha_k (\psi_m)_{j,k}(x)\right\|^2 \leq C_2 \sum_{k\in\mathbb{Z}} |\alpha_k|^2 \qquad (C.7)$$

and

$$\tilde{C}_1 \sum_{k\in\mathbb{Z}} |\tilde{\alpha}_k|^2 \leq \left\|\sum_{k\in\mathbb{Z}} \tilde{\alpha}_k (\tilde{\psi}_m)_{j,k}(x)\right\|^2 \leq \tilde{C}_2 \sum_{k\in\mathbb{Z}} |\tilde{\alpha}_k|^2, \qquad (C.8)$$

and the strictly positive constants $C_1, C_2, \tilde{C}_1$ and $\tilde{C}_2$ are independent of the scale index j.

Since W_j and $\tilde{W}_{j'}$ are orthogonal when $j \neq j'$, we have

$$\langle (\psi_m)_{j,k}, (\tilde{\psi}_m)_{j',k'} \rangle = 0 \qquad (C.9)$$

for $j \neq j'$, and from (C.5) and (C.6) we see that

$$\sum_{l\in\mathbb{Z}} \overline{\hat{\psi}_m(\omega + 2l\pi)} \, \hat{\tilde{\psi}}_m(\omega + 2l\pi) = 1. \qquad (C.10)$$

All of this means that ψ_m and $\tilde{\psi}_m$ satisfy the duality relation

$$\langle (\psi_m)_{j,k}, (\tilde{\psi}_m)_{j',k'} \rangle = \delta_{j,j'} \, \delta_{k,k'}. \qquad (C.11)$$

We have now shown that the systems $\{(\psi_m)_{j,k}\}$ and $\{(\tilde{\psi}_m)_{j,k}\}$ are biorthogonal and that they are, respectively, unconditional bases for the spaces W_j and $\tilde{W}_j$. Thus for any $f \in L^2(\mathbb{R})$

$$\Delta_j f = \sum_{k\in\mathbb{Z}} \langle f, \tilde{\psi}_{j,k} \rangle \psi_{j,k} = \sum_{k\in\mathbb{Z}} \langle f, (\tilde{\psi}_m)_{j,k} \rangle (\psi_m)_{j,k}, \qquad (C.12)$$

and

$$\tilde{\Delta}_j f = \sum_{k\in\mathbb{Z}} \langle f, \psi_{j,k} \rangle \tilde{\psi}_{j,k} = \sum_{k\in\mathbb{Z}} \langle f, (\psi_m)_{j,k} \rangle (\tilde{\psi}_m)_{j,k}, \qquad (C.13)$$

where Δ_j and $\tilde{\Delta}_j$ denote the projections on W_j and $\tilde{W}_j$.

Finally, we wish to show that $\{(\psi_m)_{j,k}\}$ and $\{(\tilde{\psi}_m)_{j,k}\}$ are unconditional bases for all of $L^2(\mathbb{R})$, and for this we need the following equivalences:

$$\|f\|^2 \sim \sum_{j,k\in\mathbb{Z}} |\langle f, (\psi_m)_{j,k} \rangle|^2 \sim \sum_{j,k\in\mathbb{Z}} |\langle f, (\tilde{\psi}_m)_{j,k} \rangle|^2. \qquad (C.14)$$

But (C.7) and (C.8) imply that

$$\sum_{j,k\in\mathbb{Z}} |\langle f, (\tilde{\psi}_m)_{j,k}\rangle|^2 \sim \sum_{j\in\mathbb{Z}} \|\Delta_j f\|^2 \qquad (C.15)$$

and

$$\sum_{j,k\in\mathbb{Z}} |\langle f, (\psi_m)_{j,k}\rangle|^2 \sim \sum_{j\in\mathbb{Z}} \|\tilde{\Delta}_j f\|^2, \qquad (C.16)$$

and the assumption that $\{\psi_{j,k}\}$ and $\{\tilde{\psi}_{j,k}\}$ are unconditional bases for $L^2(\mathbb{R})$ means that

$$\|f\|^2 \sim \sum_{j\in\mathbb{Z}} \|\Delta_j(f)\|^2 \sim \sum_{j\in\mathbb{Z}} \|\tilde{\Delta}_j(f)\|^2. \qquad (C.17)$$

Together (C.15), (C.16), and (C.17) imply (C.14) and finish the proof. □

REMARK

This expanded class of biorthogonal wavelets presents a problem in practice. In the algorithms, one must apply the filters associated with the wavelets ψ_m and $\tilde{\psi}_m$, which means that one must compute with the coefficients. At least one of these filters will have an infinite impulse response, and a truncation problem arises either in the decomposition or in the reconstruction.

We are now going to apply Proposition C.1 to the system generated by ψ_b.

Corollary C.1 *The family* $\{2^{j/2}\psi_b(2^j x - k)\}_{j,k\in\mathbb{Z}}$ *is an unconditional basis for* $L^2(\mathbb{R})$.

Proof. From what has been shown, it is sufficient to find a function ψ, which is derived from Theorem 4.1, and a regular function b, which does not vanish, such that $\hat{\psi}_b(\omega) = b(\omega)\hat{\psi}(\omega)$.

We begin by writing

$$\hat{\psi}(2\omega) = e^{-i\omega}\overline{\tilde{m}_0(\omega + \pi)} \prod_{k=1}^{+\infty} m_0(2^{-k}\omega) \qquad (C.18)$$

with

$$m_0(\omega) = \left|\frac{1+e^{i\omega}}{2}\right|^2. \qquad (C.19)$$

Using (C.2), we see that

$$\hat{\psi}_b(2\omega) = -\left(\frac{1-e^{-i\omega}}{2}\right)^2 \hat{\Delta}(\omega) = b(2\omega)\hat{\psi}(2\omega),$$

and hence that
$$b(2\omega)\,e^{-i\omega}\,\overline{\tilde{m}_0(\omega+\pi)} = -\left(\frac{1-e^{-i\omega}}{2}\right)^2$$
and
$$\overline{\tilde{m}_0(\omega+\pi)} = \frac{1}{b(2\omega)}\left|\frac{1-e^{i\omega}}{2}\right|^2. \tag{C.20}$$
Substituting (C.20) in the duality relation (C.4) shows that
$$\frac{1}{b(2\omega)}\left[\left|\frac{1-e^{i\omega}}{2}\right|^4 + \left|\frac{1+e^{i\omega}}{2}\right|^4\right] = 1,$$
from which we get
$$b(\omega) = \frac{3+\cos\omega}{4}. \tag{C.21}$$
Clearly, b is regular and non-vanishing, and
$$\tilde{m}_0(\omega) = \frac{4}{3+\cos(2\omega)}\left|\frac{1+e^{i\omega}}{2}\right|^2 = \frac{1}{b(2\omega)}\,m_0(\omega). \tag{C.22}$$

Starting with $m_0(\omega) = \left|\frac{1+e^{i\omega}}{2}\right|^2$, we have found a 2π-periodic, $\mathcal{C}^\infty$ function $\tilde{m}_0$ such that m_0 and $\tilde{m}_0$ satisfy the duality relation (C.4). We know that $|\hat{\varphi}(\omega)| \leq C(1+|\omega|)^{-2}$ because φ is the triangle function. To finish the proof, we need to verify that $\widehat{\tilde{\varphi}}$ decays sufficiently fast so that Theorem 4.1 applies. This follows directly from Proposition 3.3 by taking $p(\omega) = e^{-i\omega}/b(2\omega)$. In this case, $|p(\omega)| \leq 2$, $B_j = 2^j$, and $b_j = 1$ for all j. Then by Proposition 3.3,
$$|\widehat{\tilde{\varphi}}(\omega)| = \left|\prod_{k=1}^{+\infty} \tilde{m}_0(2^{-k}\omega)\right| \leq C(1+|\omega|)^{-1}. \tag{C.23}$$
(It is possible to derive a better decay estimate by examining the products $\left|\prod_{k=1}^n b(2^k\omega)\right|$.)

This proves that Proposition C.1 is applicable, and hence that $\{(\psi_b)_{j,k}\}_{j,k\in\mathbb{Z}}$ is an unconditional basis for $L^2(\mathbb{R})$. The dual basis $\{(\tilde{\psi}_b)_{j,k}\}_{j,k\in\mathbb{Z}}$ is given by (C.6), so

$$\begin{aligned}\widehat{\tilde{\psi}}_b(\omega) &= \frac{4}{3+\cos\omega}\,e^{-i\omega/2}\left|\frac{1-e^{i\omega/2}}{2}\right|^2 \prod_{k=2}^{+\infty} \tilde{m}_0(2^{-k}\omega) \\ &= \hat{\psi}_b(\omega) \prod_{k=0}^{+\infty} \frac{4}{3+\cos(2^{-k}\omega)}.\end{aligned}$$

□

It is seen from this last equation that $\tilde{\psi}_b$ does not have compact support. If we examine the pyramid algorithms associated with these wavelets, the filters one uses, in the case where $\tilde{\psi}_b$ is the analyzing wavelet, are as follows:
- In the decomposition, $\overline{\tilde{m}_0(\omega)} = \tilde{m}_0(\omega)$ for the approximations and $\overline{\tilde{m}_{1,b}(\omega)} = \frac{-4}{3+\cos(2\omega)}\left(\frac{1-e^{i\omega}}{2}\right)^2$ for the details.
- For the reconstruction, $m_0(\omega)$ and $m_{1,b}(\omega) = -\left(\frac{1-e^{-i\omega}}{2}\right)^2$.

It is only in the last step that the filters have finite impulse response, and this is reasonable since the functions used for the synthesis have compact support.

Before showing that the family $\{(\psi_b)_{j,k}\}_{j,k\in\mathbb{Z}}$ is also an unconditional basis for $H^1(\mathbb{R})$, we briefly examine other possible choices for the wavelet ψ.

Notice first, that a choice very close to ψ_b, at least in appearance, is ψ_c defined by

$$\psi_c(x) = \Delta(2x) - \frac{1}{2}\Delta(2x-1) - \frac{1}{2}\Delta(2x+1). \qquad (C.24)$$

However, if we use the same reasoning as in the case of ψ_b, we find that $c(\omega) = -\frac{1+e^{i\omega}}{2}$. Since $c(\pi) = 0$ it is clear that this does not work. (It is also possible to show directly that the family $\{(\psi_c)_{j,k}\}_{j,k\in\mathbb{Z}}$ cannot be an unconditional basis for $L^2(\mathbb{R})$). This indicates that the wavelet must be symmetric about $1/2$ rather than about 0. This phenomenon is also seen in the orthonormal bases constructed by Meyer and Lemarié (see Meyer (1990)).

The case of ψ_a is more obscure. Here we find that

$$a(\omega) = -\frac{3+e^{i\omega}}{4}, \qquad (C.25)$$

which is regular and does not vanish, and that

$$\tilde{m}_0(\omega) = \overline{(a(2\omega))}^{-1}\left(\frac{1+e^{-i\omega}}{2}\right) = \left(\frac{-4}{3+e^{-2i\omega}}\right)\left(\frac{1+e^{-i\omega}}{2}\right).$$

Unfortunately the techniques used in Chapters 3 and 4 to determine the decay of $\widehat{\tilde{\varphi}}$, and thus its membership in $L^2(\mathbb{R})$, do not work in this case. In particular, we note that $\left|a\left(2^n\frac{\pi}{3}\right)\right|^2 = \left|a\left(\frac{2\pi}{3}\right)\right|^2 = \frac{7}{16}$. Consequently, $|\widehat{\tilde{\varphi}}(\omega)|$ decreases at infinity slower than $|\omega|^{1-\frac{\log(7)}{2\log 2}}$ and hence slower than $|\omega|^{-1/2}$. An alternative approach to study the properties of $\tilde{\varphi}$ is to compute the eigenvalues of the transition operator $\tilde{P}_0$ associated with $\tilde{m}_0$. A difficulty with this approach arises because $\tilde{m}_0$ is not a trigonometric polynomial and, thus, $\tilde{P}_0$

cannot be restricted to a finite dimensional space. It was finally proved by A. Cohen and I. Daubechies that the spectral radius of $\tilde{P}_0$ is strictly larger than 1 so that $\tilde{\varphi}$ fails to be in $L^2(\mathbb{R})$ and ψ_a does not generate a Riesz basis.

We now return to ψ_b, which does generate an unconditional basis for $L^2(\mathbb{R})$, and show that this function also generates an unconditional basis for $H^1(\mathbb{R})$.

Proposition C.2 *The family* $\{(\psi_b)_{j,k}\}_{j,k\in\mathbb{Z}}$ *is an unconditional basis for the Sobolev space* $H^1(\mathbb{R})$. *More precisely, we have the following equivalence:*

$$\|f'\|^2 \sim \sum_{j\in\mathbb{Z}} 2^{2j} \sum_{k\in\mathbb{Z}} |\langle f, (\tilde{\psi}_b)_{j,k}\rangle|^2. \tag{C.26}$$

Proof. Consider the partial sum $s_N(f)$ defined by

$$s_N(f) = \sum_{j=-N}^{N} \sum_{k\in\mathbb{Z}} \langle f, (\tilde{\psi}_b)_{j,k}\rangle (\psi_b)_{j,k}. \tag{C.27}$$

We know that for $f \in L^2(\mathbb{R})$, $s_N(f)$ converges to f in $L^2(\mathbb{R})$. We will show that $s_N(f)$ also converges in $H^1(\mathbb{R})$ when f is in $H^1(\mathbb{R})$.

The summation on k poses no difficulty because the operator $f \mapsto \sum_{k\in\mathbb{Z}} \langle f, (\tilde{\psi}_b)_{j,k}\rangle (\psi_b)_{j,k}$ is continuous from $L^2(\mathbb{R})$ to $H^1(\mathbb{R})$. This can be shown using arguments that are similar to those used to prove Proposition 4.2. First, the mapping $f \mapsto (\langle f, (\tilde{\psi}_b)_{j,k}\rangle)_{k\in\mathbb{Z}}$ is bounded from $L^2(\mathbb{R})$ to $l^2(\mathbb{Z})$, and

$$\sum_{k\in\mathbb{Z}} |\langle f, (\tilde{\psi}_b)_{j,k}\rangle|^2 \leq \left(\sup_{\omega\in\mathbb{R}} \sum_{l\in\mathbb{Z}} |\hat{\tilde{\psi}}_b(\omega + 2l\pi)|^2\right) \|f\|^2. \tag{C.28}$$

Next, the mapping $\{c_k\}_{k\in\mathbb{Z}} \mapsto C_j = \sum_{k\in\mathbb{Z}} c_k (\psi_b)_{j,k}$ is bounded from $l^2(\mathbb{Z})$ to $H^1(\mathbb{R})$, and

$$\|C'_j\|^2 \leq 2^j \left(\sup_{\omega\in\mathbb{R}} \sum_{l\in\mathbb{Z}} |\omega + 2l\pi|^2 |\hat{\psi}_b(\omega + 2l\pi)|^2\right) \sum_{k\in\mathbb{Z}} |c_k|^2. \tag{C.29}$$

Hence, if $f_j = \sum_{k\in\mathbb{Z}} \langle f, (\tilde{\psi}_b)_{j,k}\rangle (\psi_b)_{j,k}$,

$$\|f'_j\|^2 \leq 2^j C \|f\|^2, \tag{C.30}$$

where C does not depend on j.

The problem thus reduces to showing that

$$f' = \lim_{N \to +\infty} s'_N = \lim_{N \to +\infty} \sum_{j=-N}^{N} 2^{2j} \sum_{k \in \mathbb{Z}} \langle f, \tilde{\psi}_b(2^j x - k) \rangle \psi'_b(2^j x - k) \tag{C.31}$$

(in the sense of $L^2(\mathbb{R})$).

For this, we introduce the functions $D\psi_b$ and $I\tilde{\psi}_b$ defined by

$$\widehat{D\psi_b}(\omega) = \frac{\omega}{i} \hat{\psi}_b(\omega) \tag{C.32}$$

and

$$\widehat{I\tilde{\psi}_b}(\omega) = \frac{1}{i\omega} \hat{\tilde{\psi}}_b(\omega), \tag{C.33}$$

which are thus the derivative of $-\psi_b$ and a primitive of $\tilde{\psi}_b$.

By integrating by parts the scalar products in s'_N (which is possible since f and $I\tilde{\psi}_b$ are in $H^1(\mathbb{R})$), the series can be written as

$$s'_N = \sum_{j=-N}^{N} \sum_{k \in \mathbb{Z}} \langle f', (I\tilde{\psi}_b)_{j,k} \rangle (D\psi_b)_{j,k}. \tag{C.34}$$

We next observe that the functions $D\psi_b$ and $I\tilde{\psi}_b$ satisfy the duality relation (C.11). In fact, integrating by parts shows immediately that

$$\langle (I\tilde{\psi}_b)_{j,k}, (D\psi_b)_{j',k'} \rangle = \delta_{j,j'} \delta_{k,k'}. \tag{C.35}$$

Furthermore, by using the identity

$$\prod_{k=1}^{+\infty} \left(\frac{1 + e^{2^{-k} i\omega}}{2} \right) = e^{i\omega/2} \frac{\sin(\omega/2)}{\omega/2} = \frac{e^{i\omega} - 1}{i\omega} \tag{C.36}$$

we can write

$$\widehat{D\psi_b}(2\omega) = \frac{2\omega}{i} b(2\omega) e^{-i\omega} \overline{\tilde{m}_0(\omega + \pi)} \prod_{k=1}^{+\infty} m_0(2^{-k}\omega)$$

$$= 4b(2\omega) e^{-i\omega} \overline{\tilde{M}_0(\omega + \pi)} \prod_{k=1}^{+\infty} M_0(2^{-k}\omega)$$

and

$$\widehat{I\tilde{\psi}_b}(2\omega) = \frac{1}{2i\omega}\overline{(b(2\omega))}^{-1}e^{-i\omega}\overline{m_0(\omega+\pi)}\prod_{k=1}^{+\infty}\tilde{m}_0(2^{-k}\omega)$$

$$= (4\overline{b(2\omega)})^{-1}e^{-i\omega}\overline{M_0(\omega+\pi)}\prod_{k=1}^{+\infty}\tilde{M}_0(2^{-k}\omega),$$

where

$$M_0(\omega) = \frac{2}{1+e^{i\omega}}m_0(\omega) = \frac{1+e^{-i\omega}}{2} \qquad (C.37)$$

and

$$\tilde{M}_0(\omega) = \frac{1+e^{-i\omega}}{2}\tilde{m}_0(\omega) = \left(\frac{1+e^{-i\omega}}{2}\right)^3\left(\frac{4e^{i\omega}}{3+\cos 2\omega}\right).$$

These filters ensure (by Proposition 3.3) that the associated scaling functions $\Phi(\omega) = \prod_{k=1}^{+\infty}M_0(2^{-k}\omega)$ and $\tilde{\Phi}(\omega) = \prod_{k=1}^{+\infty}\tilde{M}_0(2^{-k}\omega)$ have sufficient decay so that Theorem 4.1 is applicable. This, in turn, means that Proposition C.1 applies and, consequently, that $D\psi_b$ and $I\tilde{\psi}_b$ generate unconditional bases. Thus the identity (C.31) is proved, and from this one can establish the equivalence

$$\|f'\|^2 \sim \sum_{j,k\in\mathbb{Z}}|\langle f',(I\tilde{\psi}_b)_{j,k}\rangle|^2 = \sum_{j,k\in\mathbb{Z}}2^{2j}|\langle f,(\tilde{\psi}_b)_{j,k}\rangle|^2.$$

□

REMARK

This proof points to a more general result: If the synthesizing wavelet ψ has enough regularity and if the analyzing wavelet $\tilde{\psi}$ has enough oscillation, then one can differentiate the first and integrate the second, up to the order of interest, and thereby generate new biorthogonal bases. We note that it is not at all possible to invert the roles of the functions ψ and $\tilde{\psi}$.

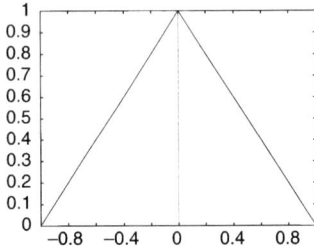

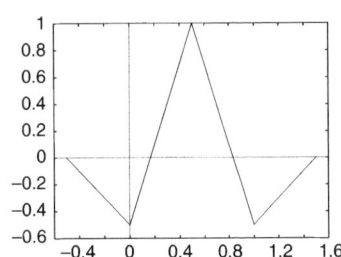

Figure C.1 *Graph of* $\Delta(x)$.

Figure C.2 *Graph of* $\psi_b(x)$.

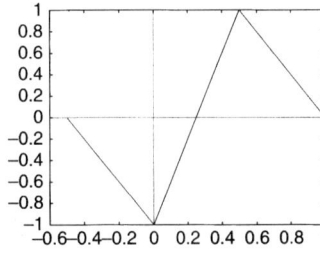

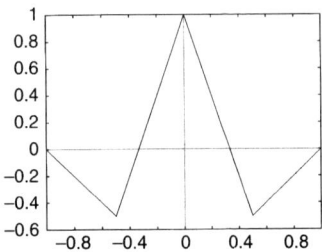

Figure C.3 *Graph of* $\psi_a(x)$.

Figure C.4 *Graph of* $\psi_c(x)$.

APPENDIX D

Notation

This appendix summarizes the notation and other conventions that are used throughout the book. We have tried to use notation that is currently 'standard' for functional analysis, harmonic analysis, and signal processing. Since certain conventions, such as the normalization of the Fourier transform, are variable, we have prepared this appendix as a quick reference.

In general, the 'signals' studied in this book are represented by functions (which are always measurable) defined on the real numbers, $\mathbb{R}$, with values in $\mathbb{C}$, the field of complex numbers, or by sequences defined on the integers, $\mathbb{Z}$, when we deal with sampled signals. $\mathbb{R}_+$ denotes the positive real numbers, and $\mathbb{N}$ denotes the integers $\{0, 1, 2, \ldots\}$.

The space or time variables are usually denoted by small Roman letters, typically x or t. The frequency variables, which appear in Fourier transforms, for example, are most often denoted by ω.

Unless otherwise indicated, the domain of integration is $\mathbb{R}$ and the domain of summation is $\mathbb{Z}$. Thus, $\int f(x)dx = \int_\mathbb{R} f(x)dx$ and $\sum s_n = \sum_\mathbb{Z} s_n$.

$L^p(\mathbb{R})$ This is the usual Lebesgue space. For $1 \leq p < \infty$, f is in $L^p(\mathbb{R})$ if and only if $\int |f(x)|^p dx < +\infty$. $f \in L^\infty(\mathbb{R})$ if and only if $|f(x)| < C$, where C is some positive constant, except perhaps for a set of measure zero.

$l^p(\mathbb{Z})$ This is the space of complex sequences $\{s_n\}_{n \in \mathbb{Z}}$ such that $\sum |s_n|^p < +\infty$. $l^\infty(\mathbb{Z})$ is the space of bounded sequences.

$H^s(\mathbb{R})$ This denotes the usual Sobolev space. For $0 \leq s < +\infty$, f is in $H^s(\mathbb{R})$ if and only if $\int |\hat{f}(\omega)|^2 (1 + |\omega|^2)^s d\omega < +\infty$.

$\mathcal{C}^\alpha(\mathbb{R})$ These are the Hölder spaces. For $0 < \alpha < 1$,

$$\mathcal{C}^\alpha(\mathbb{R}) = \left\{ f \in L^\infty(\mathbb{R}) \,\bigg|\, \sup_{x,h} \frac{|f(x+h) - f(x)|}{|h|^\alpha} < +\infty \right\}.$$

If $\alpha = n + \beta$, where n is an integer and $0 < \beta < 1$, then
$$\mathcal{C}^\alpha(\mathbb{R}) = \left\{ f \in \mathcal{C}^n(\mathbb{R}) \,\Big|\, \left(\frac{d}{dx}\right)^n f \in \mathcal{C}^\beta(\mathbb{R}) \right\}.$$

If $n \in \mathbb{N}$, $f \in \mathcal{C}^n(\mathbb{R})$ means that $\left(\frac{d}{dx}\right)^n f$ exists as an element of $L^\infty(\mathbb{R})$. $f \in \mathcal{C}^\infty(\mathbb{R})$ if and only if $f \in \mathcal{C}^n(\mathbb{R})$ for all n.

$\mathcal{S}(\mathbb{R})$ This is the Schwartz class of functions. Thus $f \in \mathcal{S}(\mathbb{R})$ if and only if $f \in \mathcal{C}^\infty(\mathbb{R})$ and $\left(\frac{d}{dx}\right)^n f(x)(1+|x|)^m$ is bounded for any $m, n \in \mathbb{N}$.

$f * g$ The convolution product of the functions f and g is defined by
$$(f * g)(x) = \int f(y) g(x-y)\, dy.$$

$s * t$ The convolution product of two sequences s_n and t_n is defined by
$$(s * t)_n = \sum_{m \in \mathbb{Z}} s_m\, t_{n-m}.$$

$\langle f, g \rangle$ The inner product of two elements f and g in $L^2(\mathbb{R})$ is defined by
$$\langle f, g \rangle = \int f(x) \overline{g(x)}\, dx.$$

$\langle s, t \rangle$ Similarly, the inner product of two elements s_n and t_n in $l^2(\mathbb{Z})$ is defined by
$$\langle s, t \rangle = \sum s_n \overline{t_n}.$$

$\delta_{0,k}$ This denotes the Kroeneker sequence defined by $\delta_{0,k} = 1$ if $k = 0$ and $\delta_{0,k} = 0$ otherwise.

$\delta_{m,n}$ $\delta_{m,n} = \delta_{0,n-m}$ for $m, n \in \mathbb{Z}$.

δ_k Where there should be no confusion, we will denote the sequence $\{\delta_{k,n}\}_{n \in \mathbb{Z}}$ by δ_k. Thus, $\{\delta_k\}_{k \in \mathbb{Z}}$ is a basis for $l^2(\mathbb{Z})$. Note that δ_k is used occasionally to denote a specific real number; this usage will be absolutely clear from the context.

$\delta(x)$ This is the distribution that consists of a Dirac mass at the point 0. When there is no chance of confusion, we sometimes denote $\delta(x - x_0)$ by $\delta(x_0)$. Thus, for example, $\delta(2l\pi)$ represents a mass of 1 at the point $2l\pi$.

NOTATION

$\mathbf{1}_E$ This is the characteristic function of the set E. Thus, $\mathbf{1}_E(x) = 1$ if $x \in E$ and $\mathbf{1}_E(x) = 0$ otherwise. In most instances E is an interval $[a, b]$.

Card$E$ This denotes the number of elements in a finite set E.

supp$f$ This is the support of f. Thus, $x \in \text{supp} f$ if and only if $f(x) \neq 0$.

$\oplus$ If A and B are vector spaces, $A \oplus B$ is the space of all vectors of the form $a + b$, $a \in A$ and $b \in B$.

$\overset{\perp}{\oplus}$ With the above notation, this means that $\langle a, b \rangle = 0$ for all $a \in A$ and $b \in B$.

$L^p[0, 2\pi]$ and $H^p[0, 2\pi]$ denote respectively the Lebesgue spaces and the Sobolev spaces for the 2π-periodic functions.

We define the Fourier transform of a function f, which will usually be an element of $L^1(\mathbb{R})$ or $L^2(\mathbb{R})$, by

$$\hat{f}(\omega) = \int f(x) e^{-i\omega x}\, dx.$$

The discrete Fourier transform of a sequence s_n is defined by

$$S(\omega) = \sum s_n e^{-in\omega}.$$

With this normalization, the n-the Fourier coefficient of the function S is given by

$$s_n = \frac{1}{2\pi} \int_0^{2\pi} S(\omega) e^{in\omega}\, d\omega.$$

In the language of filters, s_n is the 'impulse response' and S is the 'transfer function' of the system.

With our normalization and notation, Parseval's relations become

$$\frac{1}{2\pi}\langle \hat{f}, \hat{g} \rangle = \langle f, g \rangle$$

and

$$\frac{1}{2\pi} \int_0^{2\pi} S(\omega) \overline{T(\omega)}\, d\omega = \langle s, t \rangle$$

for $f, g \in L^2(\mathbb{R})$ and $s_n, t_n \in l^2(\mathbb{Z})$.

We denote the L^2 norm of $f \in L^2(\mathbb{R})$ by $\|f\| = \langle f, f \rangle^{1/2}$, and similarly, $\|s\| = \langle s, s \rangle^{1/2}$ for $s_n \in l^2(\mathbb{Z})$.

If T is an operator on $L^2(\mathbb{R})$ or $l^2(\mathbb{Z})$, we denotes its norm by $\|T\| = \sup_{\|u\|=1} \|Tu\|$.

Any two integers $j, k \in \mathbb{Z}$ define a unitary transformation on the elements of $L^2(\mathbb{R})$ that we define by

$$\psi_{j,k}(x) = 2^{j/2}\psi(2^j x - k).$$

References

Antonini, M., Barlaud, M., Daubechies, I. and Mathieu P. (1992) Image coding using wavelet transforms. *IEEE Trans. on Image Processing*, **1**, 205–220.

Burt, P. and Adelson, E. (1983) The Laplacian pyramid as a compact image code. *IEEE Trans. Comm.*, **31** 482–540.

Cavaretta, A.S., Dahmen, W. and Micchelli, C. (1991) Stationary subdivision. *Memoirs of the Amer. Math. Soc.*, **93** 1–186.

Chui, C.K. and Wang, J.Z. (1991) A cardinal spline approach to wavelets. *Proc. Amer. Math. Sco.*, **113**, 785–793.

Cohen, A. and Conze, J.P. (1992) Régularité des bases d'ondelettes et mesures ergodiques. *Rev. Math. Iberoamericana*, **8**, 351–366.

Cohen, A. and Daubechies, I. (1992) A stability criterion for biorthogonal wavelet bases and their related subband coding scheme. *Duke Math. J.*, **68**, 313–335.

Cohen, A. and Daubechies, I. (1993) Non-separable bidimensional wavelet bases. *Rev. Math. Iberoamericana*, **9**, 51–137.

Cohen, A., Daubechies, I. and Feauveau, J.C. (1992) Biorthogonal bases of compactly supported wavelets. *Comm. Pure and Appl. Math.*, **44**, 485–560.

Cohen, A., Daubechies, I. and Vial, P. (1993) Wavelets and fast wavelet transform on an interval. *Applied and Computational Harmonic Analysis*, **1**, 54–81.

Conze, J.P. and Raugi, A. (1990) Fonctions harmoniques pour un opérateur de transition. *Bull. Soc. Math. France*, **118**, 273–310.

Daubechies, I. (1988) Orthonormal bases of compactly supported wavelets. *Comm. Pure and Appl. Math.*, **41**, 909–996.

Daubechies, I. (1992) *Ten Lectures on Wavelets*. SIAM, Philadelphia.

Daubechies, I. and Lagarias, J.C. (1991) Two-scale difference equations I. Existence and global regularity of solutions. *SIAM J. Math. Anal.*, **22**, 1388–1410.

Daubechies, I. and Lagarias, J.C. (1992) Two-scale difference equations II. Infinite matrix products, local regularity and fractals. *SIAM J. Math. Anal.*, **23**, 1031–1079.

de Boor, C. (1978) *A Practical Guide to Splines*. Applied Math. Science Series nr. 27, Springer-Verlag, New York.

Deslauriers, G. and Dubuc, S. (1987) *Interpolation dyadique*. In Fractals, dimension non entières et applications, G. Cherbit ed., Masson, Paris, 44–55.

DeVore, R.A., Jawerth, B. and Lucier, B.J. (1992) Image compression through transform coding. *IEEE Trans. Inform. Theory*, **38**, 719–746.

DeVore, R.A., Jawerth, B. and Popov, V.A. (1992) Compression of wavelet decompositions. *Amer. J. of Math.*, **114**, 737–785.

DeVore, R.A. and Lorentz, G.G. (1993) *Constructive Approximation*, Springer-Verlag, New York.

Dyn, N. (1992) *Subdivision schemes in computer-aided design*. In Advances in Numerical Analysis II, Wavelets, Subdivision Algorithms and Radial Functions, W.A. Light ed., Oxford University Press, 36–104.

Eirola, T. (1992) Sobolev characterization of solutions of dilation equations. *SIAM J. Math. Anal.*, **23**, 1015–1030.

Esteban, D. and Galand, C. (1977) Application of quadrature mirror filters to split-band voice coding schemes. *Proc. IEEE Int. Conf. Acoust. Signal Speech Process.*, Hartford, Connecticut, 191–195.

Fix, G. and Strang, G. (1969) Fourier analysis of the finite element method in Ritz–Galerkin theory. *Stud. Appl. Math.*, **48** 265–273.

Gohberg, I. and Krein, M. (1969) *Introduction to the Theory of Linear Nonselfadjoint Operators*. (Translation of 1965 Russian original.) Amer. Math. Soc., Providence.

Gripenberg, G. (1993) Unconditional bases of wavelets for Sobolev spaces. *SIAM J. Math. Anal.*, **24**, 1030–1042.

Gröchenig, K. (1987) Analyse multi-échelle et bases d'ondelettes. *C. R. Acad. Sci. Paris*, **305, Série I**, 13–17.

Gröchenig, K. and Madych, W.R. (1992) Multiresolution analysis, Haar bases and selfsimilar tilings of $\mathbb{R}^n$. *IEEE Trans. Inform. Theory*, **38**, 556–568.

Grossmann, A. and Morlet, J. (1984) Decomposition of Hardy functions into square integrable wavelets of constant shape. *SIAM J. Math. Anal.*, **15**, 723–736.

Haar, A. (1910) Zur Theorie der orthogonalen Funktionen-Systeme. *Math. Ann.*, **69**, 331–371.

Jaffard, S. (1989) Exposants de Hölder en des points donnés et coéfficients d'ondelettes. *C. R. Acad. Sci. Paris*, **308, Série I**, 79–81.

Kak, A.C. and Rosenfeld, A. (1982) *Digital Picture Processing*, Academic Press, New York.

Lawton, W. (1991) Necessary and sufficient conditions for constructing orthonormal wavelet bases. *J. Math. Phys.*, **32**, 57–61.

Lemarié, P.G. (1993) Fonctions d'echelle pour les ondelettes de dimension n. *C. R. Acad. Sci. Paris*, **316–2, Série I**, 145–148.

Mallat, S. (1989) Multiresolution approximation and wavelet orthonormal bases of L^2. *Trans. Amer. Math. Soc.*, **315**, 69–88.

Meyer, Y. (1990) *Ondelettes et opérateurs, I: Ondelettes, II: Opérateurs de Calderón–Zygmund, III: Opérateurs multilinéaires.* Hermann, Paris. English translation published by the Cambridge University Press (1992).

Oppenheim, A. and Schafer, R. (1975) *Digital Signal Processing.* Prentice-Hall, New York.

Schaefer, H. (1974) *Banach Lattices and Positive Operators.* Springer-Verlag, Berlin.

Smith, M.J.T. and Barnwell, T.P. III (1986) Exact reconstruction techniques for tree-structured subband coders. *IEEE Trans. Acoust. Signal Speech Process.*, **34**, 434–441.

Tchamitchian, Ph. (1987) Biorthogonalité et théorie des opérateurs. *Rev. Math. Iberoamericana*, **3**, 163–189.

Vaidyanathan, P.P. (1992) *Multirate Systems and Filter Banks.* Prentice-Hall, Englewood Cliffs, New Jersey.

Vetterli, M. (1986) Filter banks allowing perfect reconstruction. *Signal Processing*, **10**, 219–244.

Villemoes, L. (1992) Energy moments in time and frequency for two-scale difference solutions and wavelets. *SIAM J. Math. Anal.*, **23**, 1519–1543.

Young, R.M. (1980) *An Introduction to Nonharmonic Fourier Series.* Academic Press, New York.

Zygmund, A. (1968) *Trigonometric Series, 2nd ed.*, Cambridge University Press, Cambridge, England.

Author index

Adelson, E. 7, 102, 144–146
Antonini, M. 103, 167

Barlaud, M. 103, 145, 167
Barnwell, T.P. III 3, 7, 18, 98, 101
Battle, G. 151
Burt, P. 7, 102, 144–146

Cavaretta, A.S. 70
Chui, C.K. 158
Cohen, A. vii, 4, 5, 24, 90, 97, 126, 131, 170, 177, 211, 221
Coifman, R.R. 150
Conze, J.P. 54, 57, 61, 90, 128

Dahmen, W. 70
d'Ales, J.P 177
Daubechies, I. 4, 5, 23, 24, 53, 54, 63, 64, 69, 77, 83, 95, 97, 101–103, 126, 131, 145, 148, 150, 156, 167, 168, 170, 209, 211, 212, 221
de Boor, C. 9
Deslauriers, G. 70
DeVore, R.A. 116, 167
Dubuc, S. 70
Dyn, N. 70

Eirola, T. 131
Esteban, D. 3, 7, 18, 98, 101

Fan, K. 179, 180
Feauveau, J.C. 34, 97

Fix, G. 116
Fourier, J. 1
Franklin, Ph. 7

Gabor, D. 1
Galand, C. 3, 7, 18, 98, 101
Gohberg, I. 179
Gröchnig, K. 34, 213
Grossmann, A. 199

Haar, A. 2, 7, 11,

Jawerth, B. 167

Kak, A.C. 173
Krein, M. 179

Lagarias, J. 64, 95, 148, 150
Lawton, W. 54
Lemarié, P.G. 19, 151, 220
Littlewood, J.E. 7
Lorentz, G.G. 116
Lucier, B. 167

Madych, W.R. 213
Mallat, S. 2, 3, 7, 21, 45
Mathieu, P. 103, 167
Meyer, Y. vii, 2, 4, 7, 34, 53, 63, 65, 151, 167, 168, 200, 220
Micchelli, C. 70
Morlet, J. 1, 199

Oppenheim, A. 115, 169

Paley, R.E.A.C. 7
Popov, V. 167

Raugi, A. 54, 57, 61
Rioul, O. 24
Rosenfeld, A. 173
Ryan, R.D. vii

Schafer, R. 115, 169
Smith, M.J.T. 3, 7, 18, 98, 101
Strang, G. 116

Tchamitchian, Ph. 97

Vaidyanathan, P.P. 98
Vetterli, M. 102
Vial, P. 24, 170
Villemoes, L. 131

Wang, J.Z. 158

Young, R.M. 125

Subject index

Page numbers appearing in *italic* refer to figures.

Analytic signal 199
Approximation
  see Linear approximation
  see Non-linear approximation
Approximation and compression of real images *186–97*
Approximation spaces 21
Autocorrelation function 169
  for piecewise stationary process 177–8
  TV images 172

Balian–Low theorem 63
Bessel sequences 125
Bezout's theorem 156, 209, 211
Biorthogonal wavelet bases 4, 97–9, 126
  construction of 103–25
  eigenvalues of transition operators 126–39
  enlarging the set of 216
  equivalence of norms 121–24
  examples 144–63
    see Burt and Adelson's filters
    see Spline filters
  flexible choice of filters 98
  necessary and sufficient conditions for existence of 143
  separation of analysis and synthesis 98
Burt and Adelson's filters 144–51, *153–4*
  summary table 152

Cascade algorithm 69
Compact set congruent to $[-\pi,\pi]$ modulo 2π 38, *39*
  and property (P) 39
  bivariate case 212, *213*
Compression 4, 145
  data 166–7
  experimental results 175, *186–97*
  image 166, 168, 175
  lossless 166–7
  lossy 166–7
  ratio 166, 175
Conjugate quadrature filters (CQF) 3, 16, 18
  associated with multiresolution analyses 20, 35, 37–8
  criterion to be 39
  critical exponent 79
  finite case 46
    main result 52
    roots of the filters 47–53
  finite impulse response 24, 46, 53–4
  general case 38
  not associated with multiresolution analyses 19
  transition operator P_0 56
  wavelets with compact support 53
Critical exponent 79, 84
  applications 147–51
  asymptotic behavior 94–5, *96*
  estimation of 88–92

SUBJECT INDEX

Data compression
  see Compression
Decimation 22–3, 100
Details, spaces of 21
Dilation operator 24–6
Discrete Fourier transform 21, 227

Expected error 168

Father wavelet 3, 10
Fast Fourier Transform (FFT) 24
Fast Wavelet Transform (FWT) 4
  action on oscillating signals 58–9
Filters
  see also Conjugate quadrature filters
  discrete 16
  high-pass 22
  low-pass 17, 22
Fourier analysis 1, 7
Fourier transform, defined 227
Function spaces, definitions 225–6
FWT algorithm 4, 20, 21, 24, 29
  action on oscillating signals 58, 60
  as developed by Mallat 21–3
  contrast with FFT 21, 24

Gibbs phenomenon 115, 176
Global Hölder exponent 76

Haar system 2, 11, 23, 54, 63, 83, 145, 155, 175–6
  dilated 20, 46, 59, 71, 83
Haar wavelets
  see Haar system
Hardy space 199
Heisenberg uncertainty principle 1
Hölder exponent
  see Global Hölder exponent
Hölder space 167, 225
Human visual system 168

Impulse response 16
Interpolation 23, 156, 158–61
Invariant subset 57

Karhunen–Loève basis 170, 172–4
Krein–Rutman theorem 137

Lagrangian spline 155–6
Linear approximation 165, 168–72
  biorthonormal wavelet basis 175
  experimental results *188–95*
  Haar system 175–6
  Karhunen–Loève basis 173–75
  mean square error 170
  of piecewise stationary processes 178
  orthonormal wavelet basis 170–3, 175–6
  trigonometric system 173–6
Littlewood–Paley blocks 129–31
L^p-Sobolev exponent 76
  estimation 84–8
  property (P) 84

Mapping $\omega \mapsto 2\omega$ modulo 2π
  cyclic orbits 46, 52, 59, 61
  ergodic 31
Mean square error 168–70
Multiresolution analysis 3, 7–9
  and wavelet bases 19, 37
  and wavelets with compact support 56
  discrete 19, 27
  localized 11, 12, 19, 27
    implies property (P) 39, 43
    sufficient condition to generate 45
  multivariate case 33–4, 207–14, *210*
  of $l^2(\mathbb{Z})$ 24, 26–7, 29
  of stochastic processes 4
  r-regular 64–5

SUBJECT INDEX 237

several questions about 35
Multiscale signal processing 7
Multivariate wavelets
    see Multiresolution analysis

Non-linear approximation 165, 169, 181
    experimental results *196–7*
    orthonormal wavelets 181
    trigonometric system 183
Nyquist rules 23, *23*

Orthonormal basis
    scaling function 9–10
    wavelets 10

P_0 55
Piecewise stationary process
    see Stationary process
Poisson summation 105
    Lemma 105–7
Property (P) 39, 41
    bivariate case 212
    condition for negation 46, 51
    regularity 84
    Riesz bases 45
Pyramid algorithm 7, 21–3, *22*, 100
    see also FWT algorithm
    biorthonormal bases and stability 98–9

Quadrature mirror filters 18
Quantization 167
Quasi-analytic wavelet bases
    see Wavelet bases

Reconstruction formula 23
    exact 21, 29, 35, 100–1, 210
Regularity
    convergence of subdivision algorithm 74

L^p-Sobolev exponent 76
measurement 76
scaling function 63
spatial approach 64, 95
spectral approach 64, 76
sufficient condition 77
useful in synthesis 67
wavelets 4, 63
Riesz basis
    see Unconditional basis
Riesz constants 125–6, 150

Sampling 21
Semi-frames 125
Scaling function 3, 10–12, 18, 37
    as orthonormal basis 9–10
    decay of Fourier transform 80
    defined as a distribution 54, 112
    defined by Fourier transform 19, 103
    multivariate 33–4
    regularity 63
Schwartz class of functions 226
Shift-invariant space 9
Signal-to-noise ratio 175
Sobolev regularity 167
Sobolev spaces $H^m(\mathbb{R})$ 11
    defined 225
    scaling function 37
    wavelet basis 215–23, *224*
Spline filters 151, 155–8, *162–3*
    summary table 161
Spline functions 9, 11, 98, 155
Stationary process 61, 172
    piecewise 177
Stochastic processes 165
    second order 169
Strang–Fix condition 116
Subband coding 23, 98–100, 145
    defined 100
    exact reconstruction 100–1
    examples 101

need to abandon orthogonality
101–2, 145
Subband filtering 4
Subdivision algorithms 68–71,
72–3
convergence 74
FWT algorithm 70
stationary and uniform 70
strongly convergent 70

Tensor products 34, 207
Tilings 214
fractal set *214*
Time–frequency method 1
Time–scale representation 2
Transfer function 16
Transition operators 55
action on power spectrum of
stationary process 61
eigenvalues and existence of
Riesz bases 126–39
regularity of scaling function
131
unconditional bases 99, 126

Unconditional bases 8–9, 99
see also Biorthogonal bases
for the Sobolev space $H^1(\mathbb{R})$
215–23, *224*

Wavelets 17
compact support 38, 53
approximation properties
115–6
transition operator P_0 56
energy distribution 204
Littlewood–Paley 63
Meyer 2, 63, 200
regularity 63
vanishing moments 65, 67
Wavelet bases
constructed from
multiresolution analyses 19

defined 10
for $l^2(\mathbb{Z})$ 27
on an interval 170
orthonormal 37
quasi-analytic 199–204, *205*
Wavelet transform 2, 199

DUE	RETURNED
	OCT 29 1996
MAR 12 1997	MAR 19 1997
JUL 17 1997	JUL 01 1997
DEC 29 1997	
JAN 25 1998	JUN 03 1998
5-26-98	JUN 01 1998
MAY 11 1998	DEC 17 1999
APR 20 2000	AUG 22 2000
JUL 20 2001	MAY 25 2001
JUN 12 2003	JUN 09 2003
DEC 27 2003	MAR 14 2004

BOOKS MAY BE RECALLED
BEFORE THEIR DUE DATES

Form 104